POLYMER
TOUGHENING

PLASTICS ENGINEERING

Founding Editor

Donald E. Hudgin

Professor
Clemson University
Clemson, South Carolina

Additional Volumes in Preparation

POLYMER TOUGHENING

EDITED BY

CHARLES B. ARENDS
The Dow Chemical Company
Midland, Michigan

CRC Press
Taylor & Francis Group
Boca Raton London New York

CRC Press is an imprint of the
Taylor & Francis Group, an **informa** business

CRC Press
Taylor & Francis Group
6000 Broken Sound Parkway NW, Suite 300
Boca Raton, FL 33487-2742

First issued in paperback 2019

© 1996 by Taylor & Francis Group, LLC
CRC Press is an imprint of Taylor & Francis Group, an Informa business

No claim to original U.S. Government works

ISBN-13: 978-0-8247-9474-3 (hbk)
ISBN-13: 978-0-367-40142-9 (pbk)

Visit the Taylor & Francis Web site at
http://www.taylorandfrancis.com

and the CRC Press Web site at
http://www.crcpress.com

Preface

The Dow Chemical Company has been involved in the toughening of polymers for at least 50 years. In the early 1940s Larry Amos and co-workers first introduced rubber into polystyrene. Since then, the number of polymers that have been treated to a "toughening" procedure has grown to include almost every basic polymer product. It seems as though every manufactured polymer fails under some circumstance where just a little more resistance to failure is needed to make it work. The process of producing this increased resistance to failure under mechanical stress is called "toughening." When successful, it results in better materials for converters and users. It also results in knowledge and understanding for scientists who strive to make better things and, along the way, provides the rewards of a job well done to those who are involved in the process.

Progress in toughening is slow and painful. People working with a given family of materials tend to generalize about their mechanical performance, while interpreting toughness in relationship to end-use applications. Naturally, each family of materials has unique features that distinguish it from other materials. For example, its members may be intrinsically tough but suffer from being difficult to process, or they may have that worst of all flaws: they may be expensive. On the other hand, they may be processed easily but be intrinsically brittle.

Each material can be improved for its own particular uses. But rather than treat each material as a world unto itself—independent of all other materials—it is helpful to realize that interactions between families are not rare; as we shall see in the following chapters, there is a great deal of similarity between the methods of measurement and methods of toughness enhancement. The first five chapters explore some of these similarities as background for the remaining chapters, which concentrate on toughening within specific polymer families.

In assembling this book, we had several ends in mind. We wanted to provide a forum in which similarities among otherwise disparate polymers could be addressed. This is done by examining a large portion of the polymer lines sold by one company, The Dow Chemical Company. We wanted to demonstrate the interplay between molecular structure and second-phase toughening to show how methods of manufacture are used to incorporate some of the design concepts developed in laboratory environments. By following the development of some successful products, we also gain insight into industrial research. We hope that the readers of this volume who are already working at polymer toughening will find new ways of looking at their problems through other areas of polymer research. For those who are relatively new to polymer toughening, this book should provide a basic understanding of the processes involved as well as an understanding of the dynamics of doing polymer research. Finally, data on current research are presented for those who are concerned with recent developments. The chapters are written by experts in their respective fields.

This book provides a unique perspective on polymer research in the field of toughening. It demonstrates the thought processes and basic applied research used in the development of successful polymeric products. It also helps to introduce people who are involved in the process but who may not be familiar with the methods used in related fields because they work in an atmosphere that sometimes requires a modest level of secrecy. The book is ultimately intended to provide a comprehensive overview of polymer toughening as practiced in an industrial environment. We trust that it will encourage further progress in developing new and better polymeric materials that should, in turn, bring about a richer understanding of polymers and their properties.

This volume includes contributions from many of our colleagues, whose work has been reviewed and corroborated by co-workers and represents solid achievements in polymer science and technology. The information presented here is given in good faith. Although we believe that it is valid, no warranty is implied nor should one be assumed.

Charles B. Arends

Contents

Contributors

Charles B. Arends, Ph.D.* Associate Scientist, Department of Central Research, The Dow Chemical Company, Midland Michigan

James L. Bertram, Ph.D.* Senior Associate Scientist, Thermosets Research and Development Department, The Dow Chemical Company, Freeport, Texas

Jozef Bicerano, Ph.D. Computing and Information Technology Laboratory, The Dow Chemical Company, Midland, Michigan

Matthew T. Bishop, Ph.D. Research Leader, The Dow Chemical Company, Midland, Michigan

---

*Retired

Robert A. Bubeck, Ph.D.   Associate Scientist, Central Research and Development, The Dow Chemical Company, Midland, Michigan

Bruce L. Burton, M.S.   Research Leader, Thermosets Research and Development Department, The Dow Chemical Company, Freeport, Texas

Robert A. DuBois, Ph.D.   The Dow Chemical Company, Freeport, Texas

E. I. Garcia-Meitin   Research Technologist, Analytical Sciences Polymer Center, The Dow Chemical Company, Freeport, Texas

David E. Henton, Ph.D.   Senior Associate Scientist, Plastics Department, The Dow Chemical Company, Midland, Michigan

George W. Knight*   The Dow Chemical Company, Freeport, Texas

Frederick J. McGarry   Professor of Polymer Engineering, Department of Materials Science and Engineering, Massachusetts Institute of Technology, Cambridge, Massachusetts

D. Roger Moore, Ph.D.   Project Leader, Polyurethane Product Research Laboratory, The Dow Chemical Company, Freeport, Texas

Dale M. Pickelman*   The Dow Chemical Company, Midland, Michigan

Ralph D. Priester, Jr., Ph.D.   Research Associate, Polyurethane Product Research Laboratory, The Dow Chemical Company, Freeport, Texas

---

*Retired

Jerry T. Seitz, M.S. Research Scientist, Computing and Information Technology Laboratory, The Dow Chemical Company, Midland, Michigan

David Sheih, Ph.D* The Dow Chemical Company, Freeport, Texas

Hung-Jue Sue, Ph.D.[†] The Dow Chemical Company, Freeport, Texas

Robert B. Turner* The Dow Chemical Company, Freeport, Texas

David S. Wang, Ph.D. The Dow Chemical Company, Freeport, Texas

Philip C. Yang, Ph.D. Research Leader, Dow Plastics, The Dow Chemical Company, Plaquemine, Louisiana

*Retired
†*Current affiliation:* Texas A&M University, College Station, Texas

1
Molecular Origins of Toughness in Polymers

JOZEF BICERANO and JERRY T. SEITZ
The Dow Chemical Company, Midland, Michigan

I. STRESS–STRAIN CURVES AND "TOUGHNESS"

The mechanical properties of materials are of great importance in engineering applications [1]. When a mechanical force is applied to a specimen, the deformation of the specimen is described in terms of its *stress–strain behavior*. The stress–strain behavior quantifies the *stress* (mechanical load) σ required to achieve a certain amount of *strain* (deformation or displacement) ϵ, as a function of ϵ and variables such as the temperature T and the strain rate $\dot{\epsilon}$.

An example of a stress–strain curve for a uniaxial tension experiment is shown in Fig. 1a. The deformation is reversed upon removal of the applied stress up to the yield point, beyond which permanent (plastic) deformation occurs. Strain hardening occurs as the ultimate elongation is ap-

1

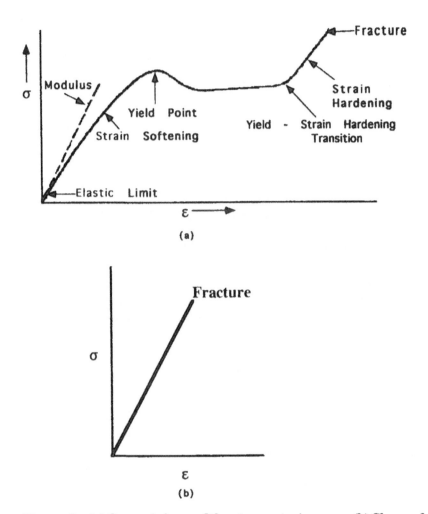

Figure 1   (a) General shape of the stress–strain curve. (b) Shape of the stress–strain curve of a very brittle material. "Toughness" is described by the total area under the stress–strain curve.

proached. Fracture occurs when the ultimate elongation is reached.

The stress has dimensions of force per unit area, that is, negative pressure. In this chapter all quantities with dimensions of stress are expressed in megapascals (MPa). The strain is always a dimensionless quantity. For a tensile deformation it is defined simply as the fractional change of the length of the specimen as a result of the deformation.

Stress–strain curves often do not show some of the features depicted in Fig. 1a. For example, for a very brittle material (Fig. 1b), they typically end abruptly in fracture after a small amount of linear elastic deformation.

Many different testing modes can be used to measure the mechanical properties of polymers. Uniaxial tension, uniaxial compression, plane strain compression, and simple shear, are among the most important testing modes. Each testing mode creates a different stress state along the three principal axes of a specimen during deformation. Several general types of stress–strain behavior are exhibited by specimens, depending on intrinsic material properties, the preparation and processing conditions of specimens, and the test conditions. See Fig. 1 for two examples. See standard references [2–6] for more detailed discussions of stress–strain curves.

There are significant differences between the *small-strain* (i.e., small deformation) and the *large-strain* behavior of polymers. Small-strain behavior will be discussed in Section II. It is mainly described by the moduli (or compliances) and Poisson's ratio. Large-strain behavior will be discussed in Section III. It refers to failure mechanisms observed in specimens, such as brittle fracture, shear yielding, and crazing, resulting in either their complete breakage or a catastrophic deterioration of their mechanical properties. In Section IV methods and computer programs that attempt to treat the mechanical properties of polymers in great atomistic detail will be discussed.

Within the context of fracture mechanics [2,6], the *tough-ness* of a specimen refers to the total amount of energy required to cause failure, that is, the total area under the stress–strain curve. For example, the specimen whose stress–strain behavior is shown in Fig. 1a is "tougher" than the specimen whose behavior is shown in Fig. 1b. Toughness is, in general, highly desirable. It can only be defined precisely for the behavior of a given specimen under a given set of test conditions. When a reference is made to the toughness of a polymer or other type of material rather than the toughness of a specimen, this describes the *statistical average* of the stress–strain behavior of a set of specimens of the material under a precisely defined set of test conditions.

Toughness is the net result of the superposition of the effects of many factors:

1. Molecular factors related to the nature of the materials. Attempts have been made to correlate these factors by means of a set of new quantitative structure–property relationships, which will be summarized later. These empirical and semiempirical relationships relate the observed behavior to the structure of the repeat unit of a polymer.
2. Chemical crosslinks, which are a special type of molecular feature that can also affect the toughness in a significant manner, as will be discussed later.
3. Effects of thermal history.
4. Effects of supramolecular organization (discussed later), as manifested by semicrystallinity, and/or other types of phase separation (as in block copolymers, immiscible polymers, and polymers reinforced with particulates or fibers).
5. Effects of anisotropy (orientation).
6. Effects of the temperature and the deformation rate during testing.
7. Effects of the mode of deformation. For example, at a given temperature and deformation rate, the proclivity to fail in

a brittle manner (i.e., not to be "tough") is much greater in plane strain tension than it is in simple shear. These effects will not be discussed in detail, since they relate to differences between the stress states imposed on the specimens as a result of the external conditions, and not to the molecular origins of toughness.

8. Effects of variations in the specimen geometry. For example, thick (bulk) specimens are more likely to fail in a brittle manner than thin films, because of the triaxial stresses created as a result of the substantial thickness of a bulk specimen. Since these effects are mainly related to the stress states created by the geometry, instead of being related to the molecular origins of toughness, they will not be discussed in detail.

9. Effects of factors related to the fabrication parameters and/ or end-use conditions. For example, specimens manufactured by injection molding often manifest some anisotropy, which affects their mechanical properties. Defects incorporated in specimens during processing or use can act as critical flaws that can cause or accelerate fracture. Exposure to harsh environments (such as oxidizing molecules) during use can cause specimens to crack. These types of factors also fall outside the main focus of this chapter and therefore will not be discussed in detail.

II. MECHANICAL BEHAVIOR AT SMALL DEFORMATIONS

A. Definitions and Phenomenology

The *modulus* is the most important small-strain mechanical property. It is the key indicator of the "stiffness" or "rigidity" of specimens made from a material. It quantifies the resistance of the specimens to mechanical deformation, in the limit of infinitesimally small deformation. There are three major types of moduli. The *bulk modulus B* is the resistance of a specimen to isotropic compression (pressure). The *Young's modulus E* is its resistance to uniaxial tension (being stretched). The *shear modulus G* is its resistance to simple shear deformation (being twisted).

Each type of modulus is defined in terms of the stresses required to deform a specimen by a strain of ϵ, in the limit of an infinitesimally small deformation of the type quantified by that modulus. For example, Young's modulus is defined by Eq. (1), in the limit of $\epsilon \to 0$ under uniaxial tension. This equation shows that the stress σ required to achieve a small strain of ϵ under uniaxial tension is proportional to E.

$$\sigma = E \cdot \epsilon$$

As a rule of thumb, the "stiffer" or "more rigid" an uncrosslinked material is, the greater is the resistance of its specimens to any type of deformation. Consequently, the larger the moduli, the less specimens "comply" with a deformation. For example, a pendant weight causes much more "creep" (extension with time, as a result of the force exerted by the hanging weight) when hung at the end of a low-modulus fiber than when hung at the end of a high-modulus fiber. The *compliance* is the reciprocal of the modulus. It indicates the extent to which the specimens of a given material are expected to comply with a deformation. The *shear compliance J*, the *bulk compliance* (or *compressibility*) κ and the *tensile compliance D* are defined as follows:

$$J \equiv \frac{1}{G} \tag{2}$$

$$\kappa \equiv \frac{1}{B} \tag{3}$$

$$D \equiv \frac{1}{E} \tag{4}$$

Poisson's ratio ν is defined by Eq. (5) for an isotropic (unoriented) specimen. It describes the effect of the application of a deformation (strain) in one direction (i.e., along the x axis) on the dimensions of the specimen along the other two directions (i.e., the y and z axes) perpendicular to the direction of the applied deformation. The fractional change of volume dV/V of the specimen is given by Eq. (6) in terms of the strains $d\epsilon_x$, $d\epsilon_y$, and $d\epsilon_z$ along the three axes. If $\nu = 0.5$, the strains along the y and z axes will each be opposite in sign and of exactly half the magnitude of the strain applied along the x axis, so that the total volume of the specimen will not change. The value of ν is very close to 0.5 for rubbery polymers. When $\nu < 0.5$, as in glassy polymers, the strains along the y and z axes will not be sufficient to counter the strain applied along the x axis, so that the total volume of the specimen will change.

$$\nu \equiv -\frac{d\epsilon_y}{d\epsilon_x} = -\frac{d\epsilon_z}{d\epsilon_x} \tag{5}$$

$$\frac{dV}{V} = d\epsilon_x + d\epsilon_y + d\epsilon_z \tag{6}$$

The value of ν provides the fundamental relationships given by Eq. (7) between moduli. It is thus only necessary to know one of the moduli, and the value of ν, to estimate the remaining two moduli by using Eq. (7), and all three compliances by using Eqs. (2), (3), and (4).

$$E = 2(1 + \nu)G = 3(1 - 2\nu)B \tag{7}$$

E, G, B and ν are functions of both the temperature and the frequency (rate) of measurement. They are often treated as complex (dynamic) properties. The real portion quantifies the energy that is reversibly stored by the "elastic" component of the deformation. The imaginary portion quantifies the energy lost (i.e., dissipated) by the "viscous" component of the deformation. For example, Eqs. (8) and (9) define the complex Young's modulus E^*, its real and imaginary components E' and E'', and the mechanical loss tangent $\tan \delta_E$ under uniaxial tension.

$$E^* \equiv E' - i \cdot E'' \tag{8}$$

$$\tan \delta_E \equiv \frac{E''}{E'} \tag{9}$$

Similar equations are also used to define the complex bulk modulus B^*, the complex shear modulus G^*, and the complex Poisson's ratio ν^*, in terms of their elastic and viscous components. The physical mechanism giving rise to the viscous portion of the mechanical properties is often called "damping" or "internal friction" and has important implications in terms of the performance of materials [7–14].

The observed moduli increase with increasing strain rate; however, the effect is usually small, and not greater than 10% over the entire range of deformation rates used in mechanical testing. Strain rate has a more significant effect on the large-strain behavior (preferred failure mechanism).

There are fundamental interrelationships, as well as significant differences, between physical factors determining the elastic and viscous components of the three types of mechanical moduli. The definitive identification of these interrelationships and differences and their embodiment in simple and reliable predictive equations are areas of ongoing research in fundamental polymer physics. We will therefore only deal with the real-valued properties E, G, B, and ν which are equivalent to the elastic (real) components E', G', B', and ν' of the corresponding complex quantities. Furthermore, only the temperature dependences of these properties will be discussed. The values calculated for these properties will therefore correspond to measurements made under "typical" testing conditions, with commonly used strain rates for each type of test.

The moduli are approximately proportional both to the strengths of the links between the atoms in a material and to the number of links per unit of cross-sectional area. (The use of the term "link" instead of "bond" is deliberate, to encompass both the covalent chemical bonds in polymer chains and the nonbonded interchain attractions such as hydrogen bonds and van der Waals interactions.) Each "link" can be viewed as a spring, with a certain value of its "spring constant" or "force constant." The moduli of a polymer (i.e., the rigidities of macroscopic specimens made from the polymer) thus generally increase with increasing chain stiffness and with increasing cohesive energy density. When a stress is applied, the weakest links (i.e., the nonbonded interchain interactions) deform much more easily than the strong covalent bonds along the individual chains. The "network" of nonbonded interchain interactions therefore plays an especially crucial role in determining the magnitudes of the moduli of a polymer.

E and G are both three to four orders of magnitude lower in rubbery polymers than in glassy polymers as a result of the breakdown of most of the interchain "links" at the glass transition. Low but nonzero values of E and G remain, however, in the "rubbery plateau" region above T_g, as a result of the restoring forces of entropic origin which oppose the deformation of the entangled polymer chains. Although most of the translational constraints to local motion disappear at T_g, a volumetric deformation under pressure does not invoke translational constraints. The reduction of B at the glass transition is therefore much less drastic than the reductions of E and G. Consequently, although all three moduli are the same order of magnitude below T_g, E and G are the same order of magnitude but B is much larger than both E and G above T_g.

The general trends for the $\nu(T)$ of amorphous polymers [15] can be summarized as follows:

1. For glassy polymers below (T_g − 20K), $\nu(T)$ typically ranges from 0.32 to 0.44. It decreases with increasing

cross-sectional area of the polymer chains. It is largest (0.40 to 0.44) for very "thin" polymer chains, and smallest (0.32 to 0.36) for very "fat" polymer chains containing large side groups. (Throughout this chapter the convention established by the American Physical Society that xK means "x degrees Kelvin" will be used consistently.)

2. $\nu(T)$ increases very slowly up to about 20K below T_g. This increase is essentially monotonic. The details of the usually small variations from a simple and smooth increase are mainly caused by factors, such as the dynamics of the polymer (e.g., its secondary relaxations) and the thermal history of a given specimen, that cannot be easily accounted for by simple and general structure–property relationships.

3. $\nu(T)$ approaches 0.5 rapidly as the specimen softens rapidly between $T = (T_g - 20K)$ and $T = T_g$.

4. $\nu(T)$ is very close to 0.5 for $T > T_g$. It is, however, always very slightly lower than 0.5, and is usually between 0.4999 and 0.5. There is a physical reason why $\nu(T)$ is not exactly equal to 0.5 for $T > T_g$. If $\nu(T)$ were exactly equal to 0.5, $B(T)$ would become infinite according to Eq. (7) since $E(T) > 0$, and the polymer would become totally incompressible above T_g according to Eq. (7), in contradiction to the observed behavior.

B. Glassy Polymers

Because of the great practical importance of the small-strain mechanical properties of polymers, many different quantitative structure–property relationships have been developed for these properties. In our practical work we have found the equations developed by Seitz [15] to have the greatest utility. Consequently, these equations have been implemented in a software package [16]. The input parameters needed to use them for a polymer of arbitrary structure can all be estimated by using correlations developed in a research monograph by

Bicerano [17] and also incorporated into the software package. The following relationships [15] are valid for the small-strain mechanical properties, up to $(T_g - 20K)$, of isotropic amorphous polymers that are glassy at room temperature (i.e., have $T_g > 298K$). Equation (13) is based on a potential function model that considers the effect of a deformation on the nonbonded (interchain) interactions in a polymer, and Eqs. (10, (11), and (12) are completely empirical.

$$\nu(298K) \approx 0.513 - 3.054 \times 10^{-6} \sqrt{\frac{V_w}{l_m}} \tag{10}$$

$$\nu_0 \approx \nu(298K) - \frac{14900}{T_g}\{0.00163 + \exp[0.459(285 - T_g)]\} \tag{11}$$

$$\nu(T) \approx \nu_0 + \frac{50T}{T_g}\{0.00163 + \exp[0.459(T - T_g - 13)]\} \tag{12}$$

$$B(T) \approx 8.23333E_{coh\,1}\left[\frac{5V(0K)^4}{V(T)^5} - \frac{3V(0K)^2}{V(T)^3}\right] \tag{13}$$

V_w denotes the van der Waals volume of the polymer in units of cc/mol, l_m denotes the length of a repeat unit of the polymer *in its fully extended conformation* in centimeters (1 Å = 10^{-8} cm), T_g is the glass transition temperature, $V(T)$ is the molar volume at the temperature of interest, $V(0K)$ is the molar volume at absolute zero temperature ($T = 0K$), and $E_{coh\,1}$ is the Fedors-type cohesive energy (see Section 5B of Bicerano [17]) in J/mol. See Bicerano [17] for the definitions and significance of V_w, T_g, $V(T)$, and $E_{coh\,1}$, methods for the estimation of these properties from the molecular structure, and tables of their values for a large number of polymers.

If only the values of the properties at room temperature are of interest, Eqs. (10) and (13) can be utilized to estimate $\nu(298K)$ and $B(298K)$, respectively, and Eqs. (2), (3), (4), and (7) can then be used to estimate the compliances and the other two moduli. If the temperature dependence of the mechanical properties is of interest, Eqs. (10), (11), and (12) must all be used to estimate $\nu(T)$, for substitution into Eq. (7).

For a vinylic polymer such as polystyrene (Fig. 2), the fully extended conformation corresponds to the all-*trans* conformation, and l_m can be estimated easily and quite accurately by assuming ideal tetrahedral bonding around the carbon atoms in the chain backbone, and using simple trigonometry. The same estimate is also valid for other polymers with vinylic backbones, such as polyethylene, poly(methyl methacrylate) and poly(vinyl chloride). For polymers with complicated repeat unit backbones, an interactive software package [16] is used to estimate l_m.

The substitution of l_m in cm, $E_{coh\,1}$ in J/mol, and V_w, $V(T)$ and $V(0K)$ in cc/mol into Eqs. (10) to (13), provides the moduli in MPa. For example [17], with $T_g \approx 373K$, $l_m \approx 2.52 \times 10^{-8}$, cm, $V_w \approx 63.25$ cc/mol, $V(298K) \approx 99.1$ cc/mol, $V(0K) \approx 91.4$ cc/mol, and $E_{coh} \approx 40,310$ J/mol used as input, predicted values of $\nu(298K) \approx 0.360$, $E(298K) \approx 3000$ MPa, $G(298K) \approx 1100$ MPa, and $B(298K) \approx 3570$ MPa, are obtained for polystyrene. (The calculated moduli are rounded off to the nearest 10 MPa, since the moduli generally cannot be measured to greater accuracy.) The most reliable available experimental results are $\nu(298K) \approx 0.354$ [15], $E(298K) \approx 3300$ MPa [15], $G(298K) \approx 1100$ MPa to 1200 MPa [18], and $B(298K) \approx 3500$ MPa [19].

Equation (13), when combined with Eq. (7), can be used to estimate $E(T)$ and $G(T)$ from $T = 0K$ up to approximately $(T_g - 20K)$. It cannot be used for $T > (T_g - 20K)$, since $\nu(T) \to 0.5$, so that $E(T)$ and $G(T)$ estimated by combining Eqs. (13) and (7) rapidly approach zero, between $(T_g - 20K)$ and T_g. As mentioned earlier, the observed $E(T)$ and $G(T)$ do not approach zero as $T \to T_g$, but gradually decrease, and then rapidly drop between T_g and approximately $(T_g + 30K)$, to the much lower range for the "rubbery plateau" region, which will be discussed in Section 2.3.

Thermal history (annealing, or "physical aging" at a high temperature below T_g) can influence the small-strain mechanical properties of a glassy polymer. Typically, the excess thermodynamic quantities ("free" or excess volume and excess enthalpy) decrease with physical aging [20–22], resulting in an increase of T_g by up to several degrees [23], and in changes of the Young's modulus [23], yielding behavior [24], and therefore toughness. The magnitude of the effects of physical aging varies widely among polymers. It is larger in polymers such as bisphenol-A polycarbonate and poly(methyl methacrylate), which have pronounced secondary (sub-T_g) relaxations [12,13] (provided that physical aging takes place above the relaxation temperature), than it is for polymers such as polystyrene that do not have pronounced secondary relaxations.

Equations (7) and (10) to (13) are also useful in estimating the effect of *crosslinking* on small-strain mechanical properties of glassy polymers. Even in a highly crosslinked polymer, there are usually sufficient numbers of weak physical (nonbonded) interactions, much more likely than the chemical bonds involved in the crosslinks to distort *under a small deformation*. The effects of crosslinking on the moduli of a glassy polymer are therefore mainly caused by the effect of the change (increase) of T_g [25–33], which is an input parameter in the equations listed above, with increasing crosslinking. Crosslinking also affects the density and thus also the value of $V(T)$; however, because of the possibility of incorporation of frozen-in free volume into the polymer during crosslinking as a result of kinetic effects, the effect on $V(T)$ is not as clearcut as the effect on T_g. The net effect of crosslinking, then, is to favor a slow increase of $E(T)$ at a given $T < T_g$ as a result of the increase of T_g, with anomalies often observed because of "frozen-in" free volume.

C. Rubbery Polymers

For glassy polymers we started from uncrosslinked polymers, and then showed how the effects of crosslinking modify the small-strain mechanical properties as described by equations developed for uncrosslinked systems. In contrast, for rubbery polymers it is more useful to start from concepts developed for crosslinked polymers, and then consider the modification and adaptation of these concepts to uncrosslinked polymers.

According to the theory of rubber elasticity [34], the "equilibrium" shear modulus $G_E°(T)$, above T_g, of a polymer crosslinked beyond its gel point, is determined by the average molecular weight M_c of the chain segments between the chemical cross links. In its simplest form, the relationship between these quantities is given by Eq. (14), where $T > T_g$, $\rho(T)$ is the density at a temperature of T, and ϕ is a factor related to the average functionality of the network. The mechanism of deformation of a rubbery polymer (entropy-controlled) is completely different from the mechanism for a glassy polymer (enthalpy-controlled), resulting in an increase in the shear modulus with increasing temperature, instead of the decrease observed for glassy polymers.

$$G_E°(T) \approx \frac{\phi\rho(\text{T})RT}{M_c} \tag{14}$$

By analogy, $G_N°(T)$ (i.e., the shear modulus of an uncrosslinked polymer in the "rubbery plateau") is usually assumed to be determined by the physical interactions caused by "entanglements" between polymer chains [35]. Equation (15) can therefore be used to define an average molecular weight between these entanglements, that is, the "entanglement molecular weight" M_e. The utilization of $R \approx 8.31451$ J/mol K) for the gas constant, units of g/cc for the density ρ, and units of g/mol for M_e results in the expression of $G_N°(T)$ in MPa by Eq. (15).

$$G_N°(T) \approx \frac{\rho(T)RT}{M_e} \tag{15}$$

In a crosslinked polymer $G_E°(T)$ increases monotonically with increasing T for a given M_c, until T becomes high enough to cause the dissociation of chemical bonds. Physical interactions resulting in the rubbery plateau in an uncrosslinked polymer are much weaker, and they gradually break as the temperature is increased in this regime. Consequently, $G_N°(T)$ does not normally increase with increasing temperature in the rubbery plateau region. Equation (15) can therefore be used to predict $G_N°(T)$ at the onset temperature of the rubbery plateau [which is often roughly at $(T_g + 30K)$], but $G_N°(T)$ is *not* proportional to the absolute temperature at higher temperatures.

A rough value of M_e can be predicted by using the empirical quantitative structure–property relationship [15,17] given by Eq. (16):

$$M_e \approx 1039.7 + 1.36411 \times 10^{-23} \frac{N_{BBrot} M V_w}{l_m^3} \tag{16}$$

In Equation (16) N_{BBrot} is the number of rotational degrees of freedom of the backbone of the polymeric repeat unit as calculated by using specific rules [16,17], and M is the molecular weight per mole of repeat units. For polystyrene, $N_{BBrot} = 2$, $M = 104.1$ g/mol, $V_w \approx 63.25$ cc/mol, and $l_m \approx 2.52 \times 10^{-8}$ so that $M_e \approx 12,265$ according to Eq. (16). "Measured" values of M_e (in reality, values back-calculated from shear modulus measurements by assuming Eq. (15) to be valid) for polystyrene are 18,700 [36] and 17,851 [15]. In general, the measurement and the prediction of M_e are both subject to substantial uncertainty. For example, the commercial-grade molecular modeling program that implements these equations [16] gives a significantly lower value of l_m than the simple trigonometric estimate shown in Fig. 2. The lower value is in much better agreement with the experimental range of M_e values for the polymer. It can be estimated for polystyrene that $\rho(403K) \approx 1.015$ g/cc so that Eq. (15) gives $G_N°(403K) \approx 0.277$ MPa.

The fact that the computational estimates and the indirectly deduced "experimental" values of M_e are both subject to substantial uncertainties does not diminish the usefulness of

(a)   (b)

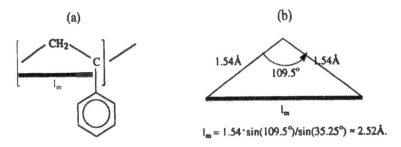

$l_m = 1.54 \cdot \sin(109.5°)/\sin(35.25°) \approx 2.52\text{Å}.$

Figure 2 Calculation of the length of a polymeric repeat unit in its fully extended conformation, with polystyrene as the example. (a) The structure of the repeat unit. The distance defined as l_m is shown. (b) Use of simple trigonometry (i.e., the sine rule), and the assumption of ideal tetrahedral bonding (bond angles of 109.5°) with a carbon–carbon bond length of 1.54 Å, to estimate that $l_m \approx 2.52$ Å = 2.52×10^{-8} cm. Molecular modeling can be used to obtain a more accurate estimate.

Eqs. (15) and (16) in predicting the general trends for the $G_N°(T)$ of rubbery polymers. Experimental values of M_e[15,36] range from 1390 for polyethylene to 110,000 for poly(t-butyl methacrylate). The largest reported value of M_e is therefore 79 times as large as the smallest reported value. It follows from Eq. (15) that the observed values [35] of $G_N°(T)$ span a range of almost two orders of magnitude. Consequently, even an error of 50% in the predicted value of M_e, which is considerably larger than the typical magnitude of the error in the prediction of this property, only results in a fairly small relative error when the entire range of $G_N°(T)$ values is considered. Trends between the $G_N°(T)$ values of members of a structurally diverse set of polymers are thus usually predicted with reasonable accuracy.

The bulk modulus $B(T)$ does not decrease nearly as drastically as $E(T)$ or $G(T)$ above T_g, but often shows a significant drop (up to a factor of 3) when T increases above T_g. Equation (17) (developed by Arends [19] by analysis of a large amount of pressure–volume–temperature data) provides a reasonable prediction of $B(T)$ for $T \geq (T_g + 30K)$. In our software package [16], Eq. (13) is used to predict B(T) for $T \leq (T_g - 20K)$. Eq. (17) is used for $T \geq (T_g + 30K)$, and an interpolation is performed for $(T_g - 20K) < T < (T_g + 30K)$.

$$B(T) \approx \frac{205V(T)/V_w}{\{[V(T)]/V_w - 1.27\}^2} - 2329 \left[\frac{V_w}{V(T)} \right] 2 \qquad (17)$$

For example, for polystyrene at $T = 403K$, the substitution of $V(403K) \approx 102.6$ cc/mol and $V_w \approx 63.25$ cc/mol into Eq. (17) gives $B(403K) \approx 1797$ MPa.

For rubbery polymers $\nu(T)$ can be estimated by rearranging Eq. (7) into the form given by Eq. (18), and then substituting the values of $B(T)$ estimated from Eq. (17) and $G(T)$ estimated from Eq. (15). Such an estimate, however, is of questionable usefulness and significance, since the $\nu(T)$ of a rubbery polymer cannot be measured to great accuracy.

$$\nu(T) \approx \frac{[3B(T) - 2G(T)]}{[6B(T) + 2G(T)]} \qquad (18)$$

For example, $G(403K) \approx 0.277$ MPa and $B(403K) \approx 1797$ MPa were predicted for polystyrene. Equation (18) then gives $\nu(403K) \approx 0.499923$. A $\nu(T)$ very close to but just below 0.5, between 0.499 and 0.499999, is typical for rubbery polymers. The approximation $\nu(T) \approx 0.5$ in Eq. (7) to relate $E(T)$ to $G_N°(T)$ for a rubbery polymer results in the following simple equation for $E(T)$:

$$E(T) \approx 3G_N°(T) \qquad (19)$$

D. Effects of Supramolecular Organization

Supramolecular organization refers to the presence of two or more distinct components or "phases" in a polymeric system, instead of a monolithic (single-phase) structure. Many types of polymeric systems manifest supramolecular organization, such as semicrystalline polymers, polymers filled with a particulate or fiber-reinforcing component, and blends, block copolymers, or graft copolymers of immiscible components. See Woodward [37] for many excellent examples.

The description of the mechanical properties of polymeric systems manifesting supramolecular organization is only at a molecular level when considering the properties of its components. The effects of the supramolecular organization itself on the mechanical properties of a polymeric system are *not* at a molecular level, but on the larger "micromechanical" scale [38–40] of the properties of a composite system constructed from given components, each component having certain properties and being present at a certain volume fraction, with the components arranged in space in a certain manner relative to each other. They are, nonetheless, worth discussing briefly because of the vast number and technological importance of multiphase polymeric systems.

Focusing on two-component systems for simplicity, the effect of supramolecular organization on the small-strain mechanical properties depends on (1) which component is the continuous (matrix) phase at a given volume fraction, (2) the shape of the discontinuous (filler) component, and (3) any anisotropy (orientation) in the spatial organization of the components. Phase continuity is discussed later. Shape and orientation are discussed in the next subsection. Note that small-strain mechanical properties are affected by the *shape* and the *orientation* of the filler component, but are not affected significantly by the *sizes* of the filler domains. This is in marked contrast to the large-strain (failure) behavior, which is discussed in Section III.

In semicrystalline polymers, and in amorphous polymers filled with a particulate or a fiber reinforcing component, the amorphous polymer component remains the matrix phase until the maximum possible packing fraction (dependent on shape) is reached for the filler phase. On the other hand, in blends, block copolymers, or graft copolymers of immiscible components, the matrix phase generally changes as a function of the volume fractions of the components, with "phase inversion" typically occurring gradually between 31% and 69% volume fraction. Such phase inversion phenomena have often been described in terms of percolation models [41–44].

See Fig. 3 for a schematic illustration. The stiff (high-modulus) phase indicated by darkly shaded regions could be the crystalline phase of a semicrystalline polymer, or a particulate filler, or the hard (polystyrene) block of a styrene–butadiene diblock copolymer. The flexible (low-modulus) component indicated by lightly shaded regions could be an amorphous polymer phase or the soft (polybutadiene) block of a styrene-butadiene diblock copolymer. Typically, phase inversion with changing volume fractions of components will occur for the diblock copolymer, but not for the other two systems in which the amorphous polymer will always remain the matrix.

The consequences of phase continuity are shown in Fig. 4. A *small* deformation (such as one used in measuring the Young's or shear moduli) mainly affects the matrix phase, with only a minimal effect on the filler phase. Consequently, in composite systems without phase inversion, the moduli

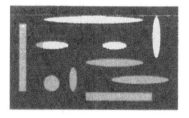
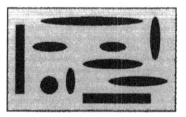

(a)   (b)

Figure 3   Schematic illustrations of (a) stiff (high-modulus, darkly shaded) matrix containing flexible (low-modulus, lightly shaded) fillers, and (b) flexible matrix containing stiff fillers.

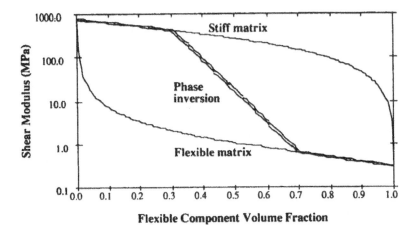

Figure 4 Schematic illustration of the effect of phase continuity on the shear modulus of a two-phase system whose components differ greatly in shear modulus. The filler particles were assumed to be spherical in calculating the stiff matrix and flexible matrix curves.

change gradually and remain close to the value for the matrix phase until a very high volume fraction of the filler is reached. On the other hand, in composite systems with phase inversion, they first change gradually and are closer to the value for component 2 (roughly over 0 to 31% by volume of component 1), then undergo relatively rapid change as a result of co-continuity of the components (31 to 69% by volume), and with the completion the reversal of the matrix and filler phases again change gradually, now being closer to those for component 1 (69 to 100% by volume).

Phenomenological models and empirical structure–property relationships are available to deal specifically with the effect of crystallinity on the mechanical properties of polymers [18,45,46]. Predictive power of wider applicability can be gained, however, by using the micromechanical methods of composite engineering [38–40], which are applicable not only to semicrystalline polymers but to *all* two-phase systems if the matrix phase is identified correctly and the shapes of the filler particles and any possible overall orientation are taken into account.

E.   Effects of Anisotropy (Orientation)

When an amorphous polymer is subjected to extensional forces at a temperature above T_g, its chains align in the direction of these extensional forces. The degree of alignment is a function of the magnitude of the applied stress. If the specimen is cooled quickly below its T_g before a significant amount of stress relaxation can take place, the orientation of the polymer chains becomes frozen in. The result is a specimen whose mechanical properties are quite different in the orientation direction and in the direction perpendicular to it, provided that there is sufficient orientation for a significant amount of chain alignment. Some anisotropy can also be induced during fabrication by methods such as injection molding [41], where polymer chains are subjected to extensional forces in addition to shear forces during melt processing.

These phenomena can be understood simply in terms of the changes in the relative numbers of chemical bonds and the much weaker nonbonded (physical) links that have to be deformed in different directions as a result of chain alignment in order to effect a macroscopic deformation of the specimen. To help visualize these molecular-level effects, see the schematic illustrations of idealized uniaxially oriented and isotropic polymers in Figs. 6a and 6b, shown in Section III.A, where they are discussed in the context of the mechanical behavior at large deformations.

Typically, the mechanical properties (such as Young's modulus and the tensile strength) improve in the direction of orientation, while becoming worse in the direction perpendicular to it. A detailed discussion of the effects of orientation was provided by Seitz [47], with amorphous polystyrene as the example. Also see the discussion by van Krevelen [18], who provided some rough empirical relationships for the effects of orientation.

The effects of orientation on the mechanical properties (such as the tensile modulus) of rubbery polymers can be large; however, since a rubbery polymer specimen normally retracts very rapidly to its original dimensions when the deformation is relaxed, the increase of its tensile modulus along the direction of chain alignment is only observed if the modulus is measured while the specimen is in the deformed state. In many industrial applications it is therefore important to know not only the usual Young's modulus of an elastomer (mea-

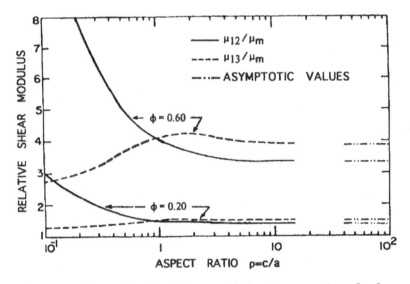

Figure 5 Example of anisotropy of elastic properties of polymers containing anisotropic fillers [40]. In this figure, μ is the shear modulus, μ_m is the shear modulus of the matrix material, and μ_{12} and μ_{13} are the moduli of a composite with uniaxially aligned fillers for shear in independent planes defined by the principal axes. Aspect ratios lower than 1 denote platelet-type fillers, an aspect ratio of 1 denotes spherical fillers, and aspect ratios greater than 1 denote fibrous fillers.

sured at a small deformation), but also its tensile modulus at large deformations (i.e., its "300% modulus" measured at an extension of 300% of the original length of the specimen).

Orientation also affects the properties of polymeric systems manifesting supramolecular organization. If the matrix phase is a polymer, its properties change as a result of orientation, and the change is reflected in the properties of the composite system. A very common example of the effects of anisotropy occurs when the filler phase consists of nonspherical particles, such as fibers or platelets [40]. The small-strain mechanical properties of polymeric systems containing a highly nonspherical filler phase depend very sensitively on the degree of nonsphericity of the filler phase (often called the "aspect ratio") and its orientation, and can be estimated by using micromechanical methods [40,48,49]. See Fig. 5 for an example.

III. MECHANICAL BEHAVIOR AT LARGE DEFORMATIONS

A. Phenomenology for Uncrosslinked Glassy Polymers

The usefulness of a polymer in many applications is largely determined by its predominant failure mechanism under the conditions of the application. Other factors being equal, a polymer whose failure requires the application of a large stress (stronger) and the absorption of more energy (tougher) will be more useful than a polymer that fails under less rigorous conditions. It is therefore very important to be able to predict the failure mechanism of polymeric specimens as a function of the structure of the polymer, the processing conditions used in the manufacture of the specimens, and the test conditions. See Ashby [1] for a broad perspective, encompassing all engineering materials, of how parameters quantifying resistance to failure can be utilized in practical guidelines for the optimum selection of materials for use in applications where specimens of various shapes are subjected to various modes of mechanical loading.

General trends for the mechanical failure of uncrosslinked amorphous specimens as a function of the material parameters and test conditions will now be summarized. Effects due to variations in the specimen geometry, differences among types of testing modes, and defects incorporated in the specimens during processing or use will not be discussed in detail, since they are not related to the fundamental molecular origins of toughness in polymers. It must be kept in mind that the predominant failure mechanism is strongly affected by these additional factors. The "material" failure properties calculated in this section are only rough indicators of the intrinsic proclivities of different polymers. Trends predicted from these material properties can be contravened by differences in the preparation and quality of the specimens of the polymers being compared.

Three general modes of failure have been identified in amorphous polymers, namely, *brittle fracture* [50], *crazing* [51–57] and *shear yielding* [2–6]. When it is not clear which mechanism is dominant, this situation is usually caused by the fact that, under the given test or application conditions, two of these mechanisms are competing, and neither mechanism is overwhelmingly favored. For example, in measurements of tensile strength or fatigue under tension, it is possible to get mixed failure modes, with crazes and shear deformation zones occurring in the same specimen. It is therefore only necessary to consider the three main mechanisms.

In *brittle fracture* (see Fig. 6 for a schematic illustration), failure occurs in a *brittle* fashion both at a *microscopic* (*local*) and a *macroscopic* (*bulk*) level. A very familiar example of brittle fracture is the shattering of ordinary (silica-based) glass (like that used in windows) when hit by a stone. Although most polymeric materials are less susceptible to brittle fracture than ordinary glass, they can all be made to undergo brittle fracture under the right set of conditions. For

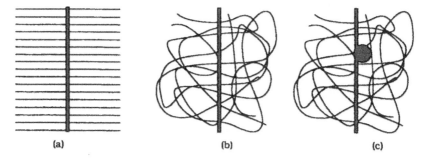

(a)   (b)   (c)

Figure 6 Schematic illustration of brittle fracture. (a) The idealized limiting case of perfectly uniaxially oriented polymer chains (horizontal lines), with a fracture surface (thick vertical line) resulting from the scission of the chain backbone bonds crossing these chains and perpendicular to them. This limit is approached, but not reached, in fracture transverse to the direction of orientation of highly oriented fibers. (b) An isotropic amorphous polymer with a typical random coil type of chain structure. Many fewer bonds cross the fracture surface (thick vertical line), and therefore many fewer bonds have to break, than was the case in the brittle fracture of a polymer whose chains are perfectly aligned and perpendicular to the fracture surface. (c) Illustration of a defect, such as a tiny dust particle (shown as a filled circle), that was incorporated into the specimen during fabrication and that can act as a stress concentrator, facilitating brittle fracture.

example, at extremely low temperatures (below $-100°C$) natural rubber can shatter and undergo brittle fracture just like glass. When a specimen undergoes brittle fracture, the bonds [whether primary (covalent) or secondary (van der Waals)] crossing the fracture surface break, and the specimen fractures as a result. There is no plastic flow during this cleavage process, since the mobilities of the subunits of the polymer are far too low. Brittle fracture has been explained in terms of (1) a defect mechanism involving a "characteristic flaw size" [2,58], and (2) considerations of the energy required for bond breaking [50]. The defect mechanism provides a better description of the fundamental physics of brittle fracture than the simple bond breakage models. On the other hand, it has proved easier to develop a simple predictive structure–property relationship for the brittle fracture stresses of polymers in terms of bond breakage.

Unlike brittle fracture, crazing (see Fig. 7 for a schematic illustration) and shear yielding (see Fig. 8 for a schematic illustration of some of the types of plastic flow processes believed to be involved [59]) both require sufficient mobility of the chain segments of a polymer for plastic flow to occur at a local (molecular) level. Both of these processes can therefore be considered to be ductile at a *local* level. On the other hand, at a *macroscopic* level, shear yielding by the homogeneous and continuous plastic deformation of a specimen is far more ductile than the heterogeneous processes of cavitation, craze nucleation, propagation and breakdown, and crack propagation, which are the stages of failure via crazing. In theories of crazing [51–54], local plastic flow processes (as quantified by the shear yield stress) and larger-scale surface separation processes (as quantified by the surface tension) must both be taken into account.

It should be clear from this discussion that there is a hierarchical scale of the molecular motions involved in the different types of fracture phenomena. A rule of thumb is that *all other factors being equal, the polymer with the most ductile mode of failure will also be the "toughest" and the most useful one.* Failure by either crazing or shear yielding is therefore to be favored over failure by brittle fracture. The desired mechanism among crazing and shear yielding is not as clear cut, and depends on additional factors such as whether specimens are notched or unnotched. Many brittle polymers are therefore "toughened" by modification (e.g., by the incorporation of rubber particles). Depending on the nature of the polymer, toughening [2,60] can occur by imparting either the ability to craze more effectively (as in polystyrene [53]) or the ability to undergo shear yielding (as in nylon [61]). Detailed discussion of

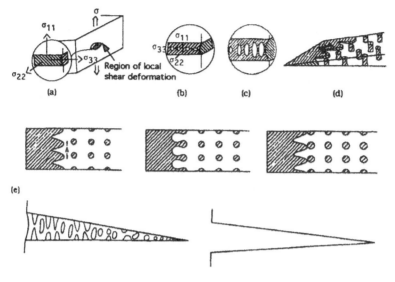

Figure 7 Schematic illustration of the stages of failure via crazing. (a) Formation of a localized plastic zone and buildup of significant lateral stresses. (b) Nucleation of *voids* in the plastic zone (very often at or near a defect close to the surface of the specimen) to relieve the triaxial constraints. (c) Further deformation of polymer ligaments between voids and coalescence of individual voids to form a void network. The columnar structures remaining between voids are craze fibrils. (d) Schematic view of the wedge of deformed polymer at the *craze tip*, showing a polymer chain about to be drawn into two different fibrils, either via chain scission or via disentanglement. (e) The advance of the craze tip (shown from the left to the right for an advance of one fibril spacing) is believed to occur via the so-called meniscus instability mechanism in which void fingers advance into the wedge of deformed polymer, leaving behind them some trailing fibrils. The void fingers and craze fibrils observed experimentally are not normally nearly as regular as those shown in this schematic illustration. (f) The difference between a craze (left) and a crack (right) is shown. As the craze tip continues to propagate, some of the craze fibrils start to break down. Once a few neighboring fibrils have broken down, a large void is formed in the craze. If the stress is high enough, these voids may grow slowly by slow fibril breakdown at their edges until a crack of critical size has formed within a craze. The crack then propagates rapidly, breaking craze fibrils as it grows, and eventually resulting in fracture. (a–e, from Ref. 53; f, from Ref. 51.)

<u>Intermolecular Shear:</u>

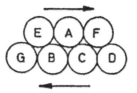

<u>Intramolecular Shear:</u>

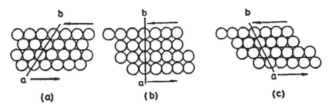

<u>Tubular Motion:</u>

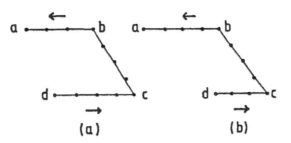

Figure 8 A highly stylized schematic illustration of some of the types of plastic flow processes believed to play an important role in shear yielding. *Intermolecular Shear:* "Atoms" E, A, and F cooperatively shear relative to atoms G, B, C, and D. *Intramolecular Shear:* The motion proceeds from (a) to (c). The difference from intermolecular shear (see above) is that the atoms connected by the line labeled a——b represent a chain "caught" at an angle to the direction of intermolecular shear, so that intermolecular shear is accompanied by a rotation of this line (chain) as it attempts to keep up with the intermolecular shear motion. *Tubular Motion:* The motion proceeds from (a) to (b). It is combined with intermolecular shear. Note the resemblance of this motion to the movements of a snake, which is approximately tubular in shape, as it glides around an obstacle. (From Ref. 59.)

the mechanisms of rubber toughening of glassy polymers is beyond the scope of this chapter.

The most common trends that determine the preferred mode of failure of a polymer are schematically depicted in Fig. 9, which shows the stresses required for brittle fracture (σ_f), crazing (σ_c), and shear yielding (σ_y) as functions of the temperature. Each failure stress will, of course, vary significantly among different polymers, so this illustration is only a rather general representation of the phenomenology of the failure of polymers.

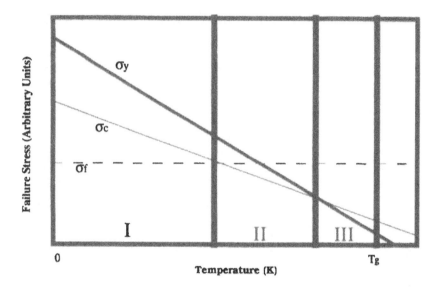

Figure 9 Schematic illustration of the stresses required for brittle fracture (σ_f), crazing (σ_c), and shear yielding (σ_y), as functions of the absolute temperature. Each roman numeral refers to the temperature regime circumscribed by the two thick vertical lines closest to it on either side. The failure stresses have been approximated by straight lines as functions of the temperature for simplicity. The experimental failure stress curves are not straight lines. Deviations from linearity are smallest (negligible for testing temperatures well below the temperature for the onset of thermal and/or thermo-oxidative degradation) for σ_f, and largest for σ_y.

As depicted in Fig. 9, σ_f generally has a very weak dependence on the temperature, since it is related to the scission of chemical bonds. The number of bonds crossing a unit cross-sectional area of a specimen decreases very slowly with thermal expansion, resulting in a very slow decrease of σ_f. The only major exception occurs when the temperature of measurement is sufficiently high for significant thermal and/or thermooxidative degradation to take place in the time scale of the test. In that case, thermal energy added to the specimen can supply most of the energy needed to break chemical bonds, and σ_f can rapidly decrease with increasing temperature.

As the temperature increases, both σ_c and σ_y decrease. The rate of decrease of σ_y with increasing temperature is usually considerably greater than the rate of decrease of σ_c because of the greater role played by plastic deformation in shear yielding. Shear yielding can be treated as a simple activated flow process, whereas crazing occurs by the superposition of plastic flow and surface separation processes.

At the lowest temperatures (regime I), σ_f is always the lowest failure stress. At an intermediate temperature range (regime II), σ_c often becomes the lowest failure stress. At the highest temperatures (regime III), σ_y often becomes the lowest failure stress.

A similar qualitative schematic illustration can also be made for the effect of the rate of deformation on the preferred failure mechanism, by relabeling the horizontal axis of Fig. 9 to indicate *decreasing* strain rate rather than *increasing* temperature. This rough equivalence between the effects of decreasing the rate of deformation and increasing the temperature is an example of the well-known time–temperature superposition [62] effects observed in viscoelastic materials. As the deformation rate is increased, motions of structural units become too slow relative to the rate of deformation, so that, for example, σ_y increases and the material becomes more brittle as a result of the suppression of plastic flow effects. This is analogous to keeping the deformation rate constant but slowing the local molecular motions by reducing the temperature. Unfortunately, simple and reasonably reliable *quantitative* structure–property relationships are available only for the temperature dependence of the failure stresses at the present time. Such relationships are therefore only intended for comparison with the results of measurements at "typical" strain rates.

The mechanical properties also depend on the ratio M_w/M_e, where M_w denotes the weight-average molecular weight. The predominant mode of failure therefore also depends on M_w. For polymers with the "typical" polydispersity of $(M_w/M_n) \approx 2.2$ to 2.5 (where M_n is the number-average molecular weight) often obtained by traditional condensation polymerization methods, an "optimum" M_w of $10M_e$ to $15M_e$ often results in the best balance of properties [15]. Polymers in this optimum range of M_w usually have chains that are long enough to approach the "high polymer" limit for the mechanical properties, but short enough for the melt viscosity to be sufficiently low not to cause great difficulty in melt processing. Methods such as anionic polymerization, which produce nearly monodisperse polymers with $(M_w/M_n) \approx 1.0$ to 1.1, can be used to synthesize polymers of much lower optimum M_w and thus of much lower melt viscosity.

When the failure stress for the mechanism requiring the smallest amount of applied stress under a given set of test or application conditions is reached, that mechanism is triggered, while competing mechanisms are not triggered. Specimens thus preferentially fail via the mechanism that requires the least amount of applied stress, with mixed failure modes or a statistical distribution of the different failure modes over a set of specimens being observed in borderline situations. Amorphous polymers therefore usually go through regimes where brittle fracture, crazing, and shear yielding are predominant (I, II, and III, respectively, in Fig. 9), with increasing temperature.

Whether a certain failure regime is observed at all and the width of the temperature range over which it is predominant vary greatly among polymers. Many polymers are so brittle that they only undergo brittle fracture below T_g, or only have regimes of brittle fracture and crazing. For example, isotropic polystyrene does not undergo shear yielding under uniaxial tension at any $T < T_g$, but only manifests brittle fracture and crazing. Furthermore, many polymers with extremely high T_g's are brittle at room temperature, and often even at elevated temperatures which that substantially lower than T_g. Great chain stiffness, which is the most important factor resulting in extremely high values of T_g [17], can make it very difficult for a polymer to dissipate a significant amount of the applied mechanical energy via plastic deformation.

Figure 9 summarizes much of the phenomenology of polymer failure, but there are some exceptions to the trends it depicts. For example, the mechanism of crazing in *high-entanglement-density polymers*, including such thermoplastics as bisphenol-A polycarbonate and poly[4,4'-isopropylidene diphenoxy di(4-phenylene)sulfone], is somewhat different from the mechanism in polymers with a low entanglement density such as polystyrene [53,54]. In polymers with a high entanglement density, the predominant failure mechanisms in regimes II and III are reversed, with specimens mainly undergoing shear yielding in regime II and crazing in regime III [63,64].

Determination of the preferred mode of failure has been shown above to require the estimation of which one of the three major types of failure stresses will be the lowest one for specimens of a given material being tested under a given set of conditions. These ideas are depicted in a flow chart in Fig. 10, assuming that a specimen is being subjected to uniaxial tension.

If a small deformation is applied to the specimen, it can break via brittle fracture when the stress σ reaches σ_f, if σ_f is the lowest failure stress. (The use of the term "small deformation" for the brittle fracture process may appear to be an oxymoron, given the fact that the specimen falls apart into two or more pieces. It is, however, an accurate usage. The *local* deformations in brittle fracture, which does not involve plastic flow

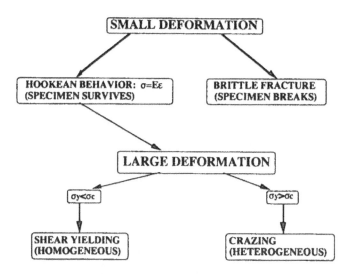

Figure 10 Flow chart for the determination of the preferred failure mechanism, assuming that a specimen is being subjected to uniaxial tension.

but only involves bond scission, are indeed small, in spite of their catastrophic effect [65,66] on the structural integrity of the specimen.) If σ_y or σ_c is the lowest failure stress, the specimen survives the small deformation, manifesting Hookean (springlike) behavior, with σ equal to the product of E and ϵ initially, and deviating from Hookean behavior with increasing ϵ. Finally, as the deformation becomes large, failure via shear yielding or crazing takes place, depending on whether σ_y or σ_c is lower.

In addition to the phenomenological models dealing with the failure of specific types of materials or with specific modes of failure, there are also some general models and computational methods that deal with universal aspects of fracture shared by all materials and all modes of failure. Such models and computational methods often provide the most appropriate description of the physical processes occurring in bulk

specimens. The statistical theory of fracture kinetics [65] for
the time-dependent fracture strength of solids is an example
of a general model attempting to incorporate the universal
aspects of fracture. The numerical methods of continuum mechanics [67,68], such as the finite element method, are leading
techniques used in design engineering to predict the mechanical properties of bulk specimens.

The bridge between the molecular and macroscopic levels
of treatment consists of the use of the quantitative structure–
property relationships to estimate the material parameters
used as input parameters in the models describing the bulk
behavior. The "intrinsic" material mechanical and thermal
properties predicted by the correlations provided in Ref. 17
can be used as input parameters in such "bulk specimen" approaches, which consider factors such as the sizes and the
shapes of the specimens. Since the predominant failure mechanism of a specimen depends strongly on many such factors
besides the intrinsic properties of the material, this type of use
of the structure–property relationships is a key step toward
developing a truly predictive capability for the preferred failure mechanisms of polymers.

B. Currently Used Quantitative Structure–Property Relationships

The brittle fracture stress σ_f is proportional to the product of
the average number of chain backbone bonds crossing a unit
area of the specimen with the average strength of a backbone
bond. This correlation was first recognized by Vincent [50]. It
was later refined by Seitz [15], who expressed it in the form
given by Eq. (20), which provides $\sigma_f(T)$ in units of MPa if l_m is
in cm and $V(T)$ is in cc/mol:

$$\sigma_f(T) \approx \frac{2.288424 \times 10^{11} \times l_m}{V(T)} \tag{20}$$

For example, for polystyrene at room temperature, $l_m \approx 2.52 \times 10^{-8}$ cm, $V(298K) \approx 99.1$ cc/mol, and $\sigma_f(298K) \approx 58$ MPa according to Eq. (20). The experimental value of σ_f for polystyrene is 41 MPa [15].

The yield stress in uniaxial tension can be estimated by assuming that it is proportional to Young's modulus [15]:

$$\sigma_y(T) \approx 0.028E(T) \tag{21}$$

For polystyrene at room temperature, $E(298K)$ was predicted to be 3000 MPa from Eqs. (13), (10), and (7). Equation (21) gives $\sigma_y(298K) \approx 84$ MPa, in reasonable agreement with the observed value [15] of 76 MPa. Even more importantly, the predicted values of $\sigma_y(298K)$ and $\sigma_f(298K)$ suggest that $\sigma_y(298K) > \sigma_f(298K)$, i.e., that brittle fracture is favored over shear yielding, for polystyrene at room temperature and commonly used strain rates. This prediction agrees with the experimental results, which show that $\sigma_y(298K) > \sigma_f(298K)$ for polystyrene.

The predicted temperature dependences [17] of the σ_y and σ_f of polystyrene are shown in Fig. 11. Note that polystyrene is thus predicted to be brittle all the way up to its glass transition.

A standard yield criterion, such as the modified von Mises criterion or the modified Tresca criterion, can be used to predict the shear yield stress in other modes of testing (such as uniaxial compression, plane strain compression, and simple shear), from the value of $\sigma_y(T)$ in uniaxial tension calculated by using Eq. (21). These two criteria are discussed in standard references [2-4] and are generally of comparable quality.

In predicting whether or not a candidate polymer will be ductile, it is recommended to err on the conservative side and not to propose that the polymer will prefer shear yielding to brittle fracture unless $\sigma_f(T) > 1.2\sigma_y(T)$, instead of using the simple criterion $\sigma_f(T) > \sigma_y(T)$ suggested by Fig. 9 for preference for shear yielding. There are several reasons for such caution. First of all, when $\sigma_f(T)$ and $\sigma_y(T)$ are very similar at

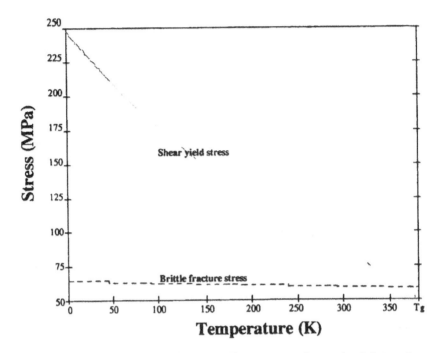

Figure 11 Calculated shear yield stress under uniaxial tension (solid line) and brittle fracture stress (dashed line) of polystyrene below the glass transition temperature. Note that the shear yield stress has a much stronger temperature dependence than the brittle fracture stress.

the temperature of interest, mixed ductile and brittle failure modes, as well as statistical variations between ductile and brittle behavior over a set of specimens, are often observed. Secondly, many important factors that were not explicitly considered in developing the quantitative structure–property relationships for the failure stresses, such as the effects of the deformation rate (as in impact phenomena where the deformation rate is extremely high), the presence of defects that act as stress concentrators, and the effects of physical aging ("ossification" of the polymer and the resulting increase in the difficulty of plastic flow processes), favor brittle fracture over shear yielding. Finally, finished parts manufactured from polymers often encounter more severe conditions than were initially anticipated.

C. An Effort to Account for the Rate Dependence of the Yield Stress

A major effort was made to account simultaneously for the temperature and the strain rate $\dot{\varepsilon}$ (i.e., the measurement frequency) dependence of the yield stress [69]. This attempt was unsuccessful in producing a reliable predictive relationship for the rate dependence. It will, nonetheless, be briefly discussed below, to help future workers in this field.

The following procedure was used to analyze data on the yield stresses of glassy polymers:

1. Begin with experimental σ_y data [70–89] obtained in various testing modes as a function of T and/or $\dot{\varepsilon}$, with emphasis on data obtained as a function of both variables in the same test on the same set of specimens.
2. Apply the modified von Mises yield criterion to convert these data into the simple shear mode, to estimate the yield stress τ_y in simple shear, and thus to provide a common working framework to compare all the data.
3. Approximate the dependence of τ_y on T and $\dot{\varepsilon}$ by using a correlation suggested by Brown [90], who combined Eyring's kinetic theory of deformation [91] with the empirical observation that τ_y is approximately proportional to the shear modulus just as $\sigma_y(T)$ in uniaxial tension is roughly proportional to Young's modulus.

$$\tau_y(T, \dot{\varepsilon}) \approx aG(T, \dot{\varepsilon}) + \left(\frac{kT}{v_s} \right) ln \left(\frac{\dot{\varepsilon}}{\dot{\varepsilon}_0} \right) \tag{22}$$

where is Boltzmann's constant and the "activation volume" [91,92] v_s, $\dot{\varepsilon}_0$, and a are material parameters. The value of a is estimated by extrapolating $\tau_y(T, \dot{\varepsilon})$ and $G(T, \dot{\varepsilon})$

to $T = 0K$ so that the second term in Eq. (22) becomes zero. Brown showed [90] that, on the average, $a \approx 0.076 \pm 0.030$ for a number of polymers.

4. Fit the experimental $\tau_y(T, \dot{\varepsilon})$ data to Eq. 22 to obtain the values of the adjustable parameters v_s, $\dot{\varepsilon}_0$, and a for as many polymers as possible.

5. Correlate the values found for v_s, $\dot{\varepsilon}_0$, and a with the structure of the polymeric repeat unit.

6. Use the correlations that will be developed for v_s, $\dot{\varepsilon}_0$, and a, in order to estimate the values of these parameters for new polymers, thus endowing Eq. (22) with predictive powers.

7. Apply the modified von Mises yield criterion to convert the predictions from τ_y in simple shear to σ_y in other testing modes.

The most important observations made as a result of this effort can be summarized as follows:

1. Contrary to expectations based on qualitative explanations commonly found in textbooks, that $\dot{\varepsilon}_0$ represents a fundamental (molecular-level) flow rate parameter, the values of $\dot{\varepsilon}_0$ show an enormous variation, not only between different polymers, but often also for data collected by different workers on the same polymer.

2. The standard treatment of the activation volume v_s for shear yielding as a constant independent of temperature for each polymer is inadequate. This oversimplification often results in large deviations between the observed values of $\tau_y(T, \dot{\varepsilon})$ and the results of fitting these data with Eq. (22).

3. The best simple functional form of $v_s(T)$ was found to be given by Eq. (23):

$$v_s(T) \approx b \exp\left(\frac{cT}{T_g}\right) \tag{23}$$

In Equation (23) b and c are fitting constants. The value of c was typically found to range from 1 to 2.75. Equation (23) implies that v_s, which is essentially the size of a typical volume element over which plastic flow takes place during shear yielding, increases exponentially with increasing temperature. This result is consistent with the increase in the mobility of the chain segments of polymers with increasing temperature.

4. The use of $v_s(T)$ as given by Eq. (23) instead of a constant value of v_s in Eq. (22) results in fits of outstanding quality between *individual sets* of experimental $\tau_y(T, \dot{\epsilon})$ data and Eq. (22).

5. It nonetheless proved to be impossible to develop reliable quantitative structure–property relationships for the parameters entering Eqs. (22) and (23) by using the available data, preventing the use of these two equations as predictive tools, for two reasons:

   a. The existence of data for σ_y simultaneously as a function of T and $\dot{\epsilon}$, over a sufficiently wide range of values of both variables, for only a few polymers.

   b. The large variations observed in some cases for the values of the adjustable parameters giving the best fit for measurements made by different workers on the same polymer.

Computer simulations [93,94] based on the kinetic theory of deformation may turn out to be a more promising approach than the development of simple structure–property relationships, in the prediction of the rate dependence of tensile deformation.

D. Efforts to Estimate the Crazing Stress

The determination of simple but reliable structure–property relationships for the crazing stress is another major challenge for future work. The complicated and heterogeneous nature of the mechanism of crazing has made the development of such relationships difficult.

The following may be some of the most promising approaches:

1. Two theories have been developed to deal with crazing under different circumstances. Future work may lead to simple structure–property relationships for the crazing stress σ_c, based on one or both of these theories.

   a. The theory of Andrews et al. [51,52] deals with environmental crazing, that is, crazing as a result of exposure, usually during the practical use of a specimen, to molecules that diffuse into the polymer and cause a deterioration of its mechanical properties. It uses a weighted arithmetic mean of the yield stress and the surface tension to calculate σ_c.

   b. The theory of Kramer [53,54] treats the mechanism of ordinary crazing in air, that is, in the absence of additional environmental agents. It uses a complicated expression, involving both the yield stress and a modified type of surface tension term, to calculate σ_c. The modified surface tension consists of the sum of two terms. The first term is the familiar surface energy related mainly to nonbonded (physical) interactions such as van der Waals interactions. The second term is included to account for the geometrically necessary loss of entanglements during craze propagation, by modifying the effective surface tension.

2. Seitz [15] has shown that the tendency to craze increases with the increasing average length of the chain segment between entanglements, while Wu [36] has correlated the crazing stress σ_c with the density of entanglements:

$$\log_{10}(\sigma_c) \approx 1.83 + 0.5 \log_{10}\left(\frac{\rho}{M_e}\right) \tag{24}$$

Both of these approaches are consistent with the physical picture provided by Kramer's theory [53,54], since they show that a more dense network of entanglements leads to a higher σ_c.

3. Kambour [55,56] has developed a correlation for the minimum stress required for the inception of crazing, in terms of the cohesive energy density, T_g, and E (or σ_y). Since failure with crazing as the primary mechanism requires the growth and propagation of crazes after their inception, this correlation describes only one of the key aspects of crazing. It is, however, a major step in the development of a complete structure–property relationship.

E. Effects of Crosslinking on the Preferred Failure Mechanism

When a polymer is crosslinked, the degrees of freedom available for motions of the chain segments is reduced as a result of the replacement of weak nonbonded (physical) interactions by much stronger chemical bonds. Large-scale plastic flow processes, such as those occurring in either shear yielding or crazing, therefore become more difficult. Consequently, when a polymer is crosslinked very densely, the usual result is embrittlement (reduction of "toughness"). At low crosslink densities, on the other hand, the effects are often relatively minor. These trends are shown for the case of styrene–divinylbenzene copolymers in Fig. 12, which was redrawn from information provided in Amos et al. [95]. Styrene is a difunctional monomer, whereas divinylbenzene is a tetrafunctional monomer. The number of repeat units between crosslinks therefore decreases (i.e., the crosslink density increases) with increasing mole fraction of divinylbenzene.

F. Effects of Supramolecular Organization

Pearson and Yee [96] studied the effect of crosslink density on the toughness of epoxy resins. They showed that the fracture

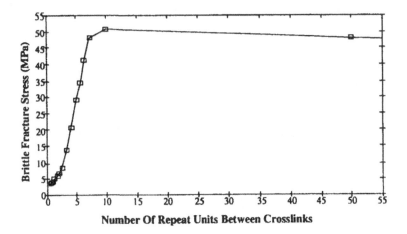

Figure 12   The effects of crosslinking on the brittle fracture stress of styrene–divinylbenzene copolymers. Note the catastrophic embrittlement at very high crosslink densities (i.e., at a very small average number of repeat units between crosslinks).

toughness of neat epoxies is very low and almost independent of the average molecular weight M_c between crosslinks in the range examined. When the epoxies were used as the matrix, and 10% by volume of an elastomer (a carboxyl-terminated copolymer of butadiene and acrylonitrile) was added as a filler, the fracture toughness of the two-phase system increased rapidly with increasing M_c. Consequently, although the unmodified epoxides were all very brittle, their "toughenability" increased rapidly with decreasing crosslink density.

   This is only one example of the great complexity of fracture processes observed in polymeric systems manifesting supramolecular organization, that is, containing two or more distinct phases. The toughness of the two-phase system could not have been predicted simply on the basis of the properties of the epoxy matrix (of similar fracture toughness in each case), the elastomer (the same in each case), and the volume fractions of the components (the same in each case).

See Refs. 2, 53, 60, and 61 for additional examples of rubber toughening.

More generally, the mechanism of failure of a multiphase system depends on the details of the morphology of the system, including the sizes of the different components and the distances between them. The situation is therefore far more complex than found for the behavior under small deformations, which normally depends on the shapes and orientations of the phases but not on the two additional variables of their sizes and distances from each other.

Furthermore, the effect of the strength of the interface (often referred to as "bonding" or "adhesion" effects) between the matrix and filler phases is also very important in determining the preferred failure mechanism of a multiphase system, whether all phases are polymers as in the examples of rubber toughening given above, or one of the phases is a particulate filler or fiber.

The importance of these additional factors, and the complications they introduce, can be seen by considering a blend and a block copolymer of the same two immiscible components, with the same overall composition in each case. It was seen earlier that the same micromechanical treatment is applicable to the behavior of the small-strain properties (i.e., moduli) of both types of systems. On the other hand, the large-deformation behavior of the two systems can be drastically different, so that current micromechanical methods are not useful in predicting the failure behavior of these or other types of multiphase systems in a general and quantitative manner. For example, the blocks in the block copolymer are chemically bonded to each other, so that the two phases are unavoidably strongly bonded [97]. On the other hand, there is no such automatically strong bonding between the two phases in a blend, whose phases in fact may have a very weak interface.

Woodward [37] is a good source on the fracture of selected examples of all types of multiphase polymeric systems, since it enables the reader to *visualize* the complex processes taking place during fracture. Tsai and Hahn [38] provide a useful discussion of the fracture of polymeric composites from a micromechanical viewpoint. Kinloch and Young [2], in addition to providing a detailed discussion of toughened multiphase plastics, also give an extensive treatment of the effects of crystallinity on fracture. Semicrystalline polymers are the only type of polymeric system manifesting supramolecular organization for which a general sequence of steps *at the molecular level* leading up to fracture can be suggested. See Fig. 13. The following discussion is based on that of Samuels [46,98–100]. In a typical semicrystalline polymer the crystalline domains are in the form of large superstructures called *spherulites*. The crystals are aligned along the radii in these spherulites, in a pattern resembling the spokes of a wheel. When the specimen is stretched uniaxially, the spherulites

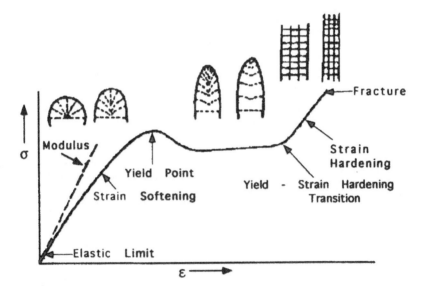

Figure 13   Schematic representation of the structural changes that occur in the crystalline domains, in different regions of the stress–strain curve of a semicrystalline polymer. (From Ref. 98.)

deform into ellipsoidal shapes, with rearrangement of the polymer chains taking place in both crystalline and amorphous domains. These rearrangements become increasingly irreversible past the yield point. The tightly interconnected extended elliptical spherulite structure is destroyed, and microfibrillar structures are formed, in the strain hardening region. Further deformation continues primarily by orientation of the polymer chains in the amorphous domains, since the crystalline domains are now at their maximum possible alignment. Finally, the structure as a whole cannot support the stress, flaws develop, and the specimen breaks.

Foams [101] are composite systems that have a *solid* phase (polymer, ceramic, or metal) and a *fluid* phase (a gas in most synthetic foams, a liquid in most natural foams). A foam has at least two components. At a macroscopic scale there are the solid and fluid phases. At a microscopic scale, the solid may have several components. For example, the solid phase of an amorphous polystyrene foam has only one component. On the other hand, the solid phase of a polyethylene or a polyurethane foam typically has two components (the crystalline and amorphous phases in polyethylene, hard and soft segment blocks in polyurethane). The solid phase of a polyurethane foam may, in fact, have even more than two components, since additional reinforcing components such as styrene–acrylonitrile copolymer or polyurea particles are often incorporated [102].

The solid is always a continuous phase in a foam. In a closed-cell foam, the fluid is isolated in the cells and is thus the discontinuous phase, although it may occupy up to 99% of the volume. In an open-cell foam, the cells are interconnected, so the fluid is co-continuous with the solid.

The mechanical properties of foams (including their toughness) are best treated not in terms of the usual composite micromechanical equations, but instead via simple mechanical models based on *beam theory* combined with scaling concepts [101]. The key parameters determining the mechanical properties of a foam at both small and large deformations are (1) the mechanical properties of the solid phase, (2) the relative density (i.e., the factor by which the density of the foam is smaller than the density of the solid from which it is made), (3) whether the cells are open or closed, and (4) the extent and nature of the anisotropy (of the individual cells and/or of their supramolecular organization) that is present in almost all foams (whether natural or synthetic).

G. Effects of Anisotropy (Orientation)

The effect of orientation is generally to "strengthen" the polymer in the direction of the orientation, and to "weaken" it in the direction perpendicular to it. This effect has a simple molecular origin, related to the alignment of polymer chains, which was discussed in Section II.E.

Consider brittle fracture (Fig. 6). Breakage of many chemical bonds along chain backbones is needed to fracture the specimen in the direction of alignment in the perfectly oriented polymer shown in Fig. 6a. Fracture perpendicular to the direction of chain alignment only requires the breakage of much weaker nonbonded (physical) interchain interactions. For a typical polymer, the chains in a macroscopic specimen are never all perfectly aligned, so that Fig. 6a depicts an idealized theoretical limit. See Adams et al. [103] for a compendium of recent research on "rigid rod" polymers, whose tensile properties can be made to approach this theoretical limit by using sophisticated fabrication methods. In the isotropic (unoriented) system shown in Fig. 6b, the average number of chemical bonds that must be broken to fracture the specimen (and thus also the brittle fracture stress) would be the same in all directions and intermediate between what would be expected for the longitudinal and transverse directions of the perfectly oriented system.

Next, consider crazing, illustrated schematically in Fig. 7. As discussed by Kinloch and Young [2], if the applied stress is parallel to the direction of orientation, then craze initiation and growth, and subsequent craze breakdown and crack extension, are both inhibited and occur at higher stress levels than they would occur at in the isotropic material. Conversely, if the applied stress is perpendicular to the direction of orientation, these events occur more readily.

Finally, consider shear yielding. As shown in Fig. 8, the plastic flow processes involved in shear yielding can be visualized [59] in terms of intermolecular shear, intramolecular shear, and tubular motions. When a specimen is oriented, intermolecular shear is suppressed in the direction of chain alignment, and tubular motions also become far more difficult, resulting in an increase in the yield stress. Conversely, intermolecular shear and tubular motions are both facilitated perpendicular to the direction of chain alignment, reducing the yield stress in that direction.

See Seitz [47] for a review of the effects of uniaxial and biaxial orientation on the fracture of polystyrene (which fails by either brittle fracture or crazing) under uniaxial tension and impact.

See Kinloch and Young [2] for additional examples of the effects of orientation on fracture, including semicrystalline polymers to which the principles discussed earlier are also applicable. The same concepts are also applicable to composite systems [38] containing unidirectional fibers, where the alignment of the fibers has an effect on the strength of the composite in different directions that is similar to the effect of the alignment of the chains in an amorphous polymer. See Gibson and Ashby [101] for the effects of anisotropy on the mechanical properties of foams.

IV. DETAILED ATOMISTIC SIMULATION METHODS

During mechanical deformation, the applied stress affects both the nonbonded interactions and the chemical bonds in a specimen, resulting in changes in the specimen geometry and ultimately causing fracture. The molecular origins of toughness can be comprehended through the manner in which the applied stress affects the structure of the specimen at this detailed atomistic level.

In recent years, with rapid increases in the speed and major improvements in the architecture of computers, detailed atomistic simulations of complex physical processes have become possible [104,105]. It is therefore reasonable to ask whether such simulations could be used to address the issue of toughness in polymers and to provide insights and predictive capabilities on a truly molecular scale. To answer this question, the natures of the physical phenomena involved in polymer failure, and the size and time scales of these phenomena, must be considered carefully.

First, consider small deformations. It is possible to simulate both crystalline systems and amorphous systems containing up to several thousand atoms, while avoiding surface (finite size) effects by using appropriate mathematical tricks. Interatomic interactions are represented by a "force field" that describes how the potential energy of the system changes as a function of the coordinates of the atoms. Bonded interactions include the energy required to stretch or to bend chemical bonds. Nonbonded interactions include the van der Waals, electrostatic, and hydrogen-bonding forces. The geometry of a model system is optimized by starting from an initial guess for atomic coordinates and finding a minimum-energy configuration by varying the coordinates iteratively. The moduli can be estimated by applying an external force of given magnitude to the model system, calculating the effect of this external force on the dimensions of the system, and utilizing the usual definitions of the linear elastic properties to estimate the moduli. Note that this "computer experiment" is conceptually quite similar to an actual measurement of the moduli.

Unfortunately, as will be shown below, the situation is far more complicated for large deformations, and little can be accomplished by using detailed atomistic simulations with the existing computational hardware and software.

Brittle fracture (Fig. 6) requires breaking chemical bonds. Chemical bonding involves electrons and is therefore inherently quantum mechanical [106,107]. Proper description of bond dissociation requires a quantum mechanical treatment and thus cannot be accomplished by using "classical" force fields in which atoms are represented as balls connected by springs subject to the forces of Newtonian mechanics. Since quantum mechanical calculations require much more computer time and memory than simulations using classical force fields, their incorporation into the routine detailed atomistic simulation of large model systems is still in the distant future.

Crazing (Fig. 7) involves the inception, propagation, and destruction of very large structural entities, such as craze fibrils, void fingers, and cracks [53,54]. The sizes of these entities are far beyond the current capabilities of even the fastest supercomputers with the most powerful memories. Furthermore, it can be seen from Fig. 7 that chain scission also occurs during failure by crazing, and as discussed earlier, chain scission (bond breakage) cannot be treated properly without taking quantum mechanical effects into account.

The typical activation volumes [69] over which plastic flow takes place during shear yielding (Fig. 8) are also quite large. Consequently, with current computer hardware, it is impossible to simulate model systems that are sufficiently large (i.e., at least two orders of magnitude larger than the typical activation volume) for the results to be statistically meaningful.

The superstructures found in multiphase polymeric systems (see Fig. 13 for semicrystalline polymers) are usually so large that it is impossible to simulate the failure behavior of such systems in realistic atomistic detail by utilizing a representative volume element of the necessary size.

Finally, the time scales involved in the "real-time" fracture of a specimen by crazing or shear yielding during a mechanical test are many orders of magnitude longer than the time scales accessible with current computational hardware and software, even when the model systems used in the simulation are much smaller than what is really needed to provide a realistic description of the important physical processes.

In conclusion, the current hardware and atomistic simulation software technology has just reached the point at which a realistic treatment of small-strain (linear elastic) mechanical properties at a fully molecular level has become a possibility. There are very few examples of even such a limited atomistic treatment of small-strain mechanical properties in the scientific literature, since the acquisition of this capability is very recent. The more ambitious goal of treating "toughness" (i.e., failure processes) at the level of a fully atomistic simulation is still in the distant future.

The limiting factor is not our understanding of the fundamental physical forces of nature that determine the bonded and nonbonded interactions of atoms. It is simply the lack of the necessary computational "horsepower" (speed, memory, algorithms of sufficient efficiency) to attempt to simulate processes occurring at size scales that are larger and time scales that are longer, often by many orders of magnitude, than what is currently accessible. Given the great ongoing improvements in the speed and memory of computers, in their architecture (i.e., the use of parallel processes in calculations whose steps can be run separately and simultaneously), and consequently in the diversity and the difficulty of the problems they are capable of addressing [108], we can only hope that the day may eventually come when it will be possible to simulate fracture processes and to understand and predict "toughness" at a truly atomistic level.

V. SUMMARY AND CONCLUSIONS

The usefulness of a polymer in many applications is determined by the predominant failure mechanism of specimens fabricated from it, under mechanical deformations. A polymer whose failure generally requires a larger amount of energy (i.e., a "tougher" polymer) is more useful than a polymer that fails under less rigorous conditions.

The molecular origins of toughness in polymeric systems were discussed in this chapter. Starting with a discussion of the much simpler mechanical properties under small deformations, the discussion then proceeded to behavior under large deformations, as described by several types of key failure mechanisms.

In amorphous uncrosslinked isotropic polymers, the three major failure mechanisms are brittle fracture, crazing, and shear yielding. In brittle fracture, failure occurs in a brittle fashion both at a microscopic (local) and a macroscopic (bulk) level. Crazing and shear yielding both require sufficient mobility of chain segments for plastic flow to occur at a local (molecular) level. Both crazing and shear yielding can thus be considered to be ductile at a local level. On the other hand, at a macroscopic level, shear yielding via a homogeneous and continuous plastic deformation of the specimen is far more ductile than the heterogeneous processes of cavitation, craze nucleation, propagation and breakdown, and crack propagation, which are the stages of failure via crazing. A general rule of thumb is that if all other factors are equal, then the polymer with the most ductile mode of failure will be the most useful one. Failure by either crazing or shear yielding is therefore to be favored over failure by brittle fracture. The desired mechanism among crazing and shear yielding is not as clear cut and depends on additional factors such as whether specimens are notched or unnotched. Failure normally occurs by the mechanism that requires application of the least amount of stress, since that mechanism is then activated while the others are not. The prediction of the predominant failure mechanism for specimens of a given material under a given set of test conditions can therefore be reduced to the task of estimating the stresses required to induce failure by each one of the three mechanisms under the prevailing set of conditions.

The fracture of crosslinked polymers, the fracture of multiphase polymeric systems (such as semicrystalline polymers; blends, block copolymers, and graft copolymers of immiscible components; and polymers containing particulate or fibrous fillers), and the effects of anisotropy (orientation), were also discussed. The behavior of multiphase polymeric systems was shown to result from the superposition of molecular factors and the effects of supramolecular organization.

Finally, the promise and the limitations of detailed computationally intensive atomistic simulations as possible new tools for investigating toughness were reviewed.

ACKNOWLEDGMENT

We thank Chuck Arends for encouragement, for helpful discussions, and for reviewing the manuscript.

GLOSSARY

A.   Terms Starting with a Lowercase Letter of the Latin Alphabet

a,b,c	Generic symbols for a constant or an adjustable parameter in a regression equation
cc	Cubic centimeter (a unit of volume, equal to cm^3)
cm	Centimeter
dV	Change in volume as a result of deformation
$d\varepsilon_x$	Deformation (strain) in the x-direction
$d\varepsilon_y$	Deformation (strain) in the y-direction
$d\varepsilon_z$	Deformation (strain) in the z-direction
f	Filler or fiber phase of a two-phase system (used as a subscript).
g	Gram (a unit of weight)
k	Boltzmann's constant
l_m	Length (end-to-end distance) of a single repeat unit of a polymer

m	Matrix phase of a two-phase system (used as a subscript)
$\tan \delta_E$	Mechanical loss tangent under the uniaxial tension mode of deformation
v_s	Activation volume for shear yielding

B. Terms Starting with a Capital Letter of the Latin Alphabet

Å	Ångstrom (a unit of length, equal to 10^{-8} cm)
B	Bulk modulus
B^*	Complex bulk modulus, with real (elastic) and imaginary (viscous) components
B'	Real (elastic) component of B^*, equivalent to B
BB	Polymer chain backbone
D	Tensile compliance (reciprocal of Young's modulus E)
E	Young's modulus
E^*	Complex Young's modulus, with real (elastic) and imaginary (viscous) components
E'	Real (elastic) component of E^*, equivalent to E
E''	Imaginary (viscous) component of E^*
$E_{coh\,1}$	Fedors-like cohesive energy calculated as described in Section 5B of Bicerano [17]
G	Shear modulus
G^*	Complex shear modulus, with real (elastic) and imaginary (viscous) components
G'	Real (elastic) component of G^*, equivalent to the second definition of G
$G_E{}^\circ$	Equilibrium shear modulus for $T > T_g$ in a polymer crosslinked beyond its gel point
$G_N{}^\circ$	Shear modulus of an uncrosslinked polymer in the rubbery plateau region, i.e., for $T \geq (T_g + 30\text{K})$ but not so high as to be in the terminal zone
J	Joule (a unit of energy)
J	Shear compliance (reciprocal of the shear modulus G)
K	Abbreviation for "degrees Kelvin"
M	Molecular weight per repeat unit
M_c	Average molecular weight between chemical crosslinks in a polymer
M_e	Entanglement molecular weight of a polymer
M_n	Number-average molecular weight of a polymer

MPa	Megapascals (a unit with dimensions of pressure, used for stresses and moduli)
M_w	Weight-average molecular weight of a polymer
N_{BBrot}	Number of rotational degrees of freedom in the backbone of a polymeric repeat unit
R	Gas constant, equal to 8.31451 J/(mol K)
T	Absolute temperature, in degrees Kelvin
T_g	Glass transition temperature, in degrees Kelvin
V	Molar volume
V_w	Van der Waals volume

C. Terms Starting with a Lowercase Letter of the Greek Alphabet

These terms are listed below in the alphabetical order of the corresponding letters of the Latin alphabet: $e \rightarrow \epsilon$, $g \rightarrow \gamma$, $k \rightarrow \kappa$, $n \rightarrow \nu$, $r \rightarrow \rho$, $s \rightarrow \sigma$, and $t \rightarrow \tau$.

ϵ	Amount of strain a specimen is subjected to, as the fractional change of its length
$\dot{\epsilon}$	Strain rate, defined as (fractional change in length of specimen)/ seconds
$\dot{\epsilon}_0$	A parameter in equation for the shear yield stress as a function of temperature and $\dot{\epsilon}$
$\dot{\gamma}$	Shear rate
κ	Compressibility (defined as the bulk compliance, i.e., reciprocal of bulk modulus B)
ν	Poisson's ratio, which relates the bulk, shear, and Young's moduli to each other
ν^*	Complex Poisson's ratio, with real (elastic) and imaginary (viscous) components
ν'	Real component of ν^*, equivalent to the static Poisson's ratio ν
ρ	Density
σ	Stress applied to a specimen during mechanical deformation
σ_c	Stress required to cause failure of a specimen by crazing
σ_f	Stress required to cause failure of a specimen by brittle fracture
σ_y	Stress required to cause failure of a specimen by shear yielding
τ	Shear stress
τ_y	Stress required to cause failure of a specimen by shear yielding under simple shear

REFERENCES

1. M. F. Ashby, *Materials Selection in Mechanical Design*, Pergamon Press, New York, 1992.
2. A. J. Kinloch and R. J. Young, *Fracture Behaviour of Polymers*, Elsevier Applied Science Publishers, London, 1983.
3. R. N. Haward, Ed., *The Physics of Glassy Polymers*, Applied Science Publishers, London, 1973.
4. W. Brostow and R. D. Corneliussen, eds., *Failure of Plastics*, Hanser Publishers, Munich, 1986.
5. N. G. McCrum, C. P. Buckley, and C. B. Bucknall, *Principles of Polymer Engineering*, Oxford University Press, New York, 1988.
6. J. G. Williams, *Fracture Mechanics of Polymers*, Ellis Horwood, West Sussex, England, 1984.
7. C. A. Wert, *J. Appl. Phys. 60*:1888 (1986).
8. A. Hiltner and E. Baer, *Polymer 15*:805 (1974).
9. R. J. Crawford and P. P. Benham, *J. Mater*. Sci. *9*:18 (1974).
10. R. J. Crawford and P. P. Benham, *J. Mater. Sci. 9*:1297 (1974).
11. R. J. Crawford and P. P. Benham, *Polymer 16*:908 (1975).
12. J. Bicerano, *J. Polym. Sci., Polym. Phys. Ed. 29*:1329 (1991).
13. J. Bicerano, *J. Polym. Sci., Polym. Phys. Ed. 29*:1345 (1991).
14. N. G. McCrum, B. E. Read, and G. Williams, *Anelastic and Dielectric Effects in Polymeric Solids*, John Wiley, New York, 1967.
15. J. T. Seitz, *J. Appl. Polym. Sci. 49*:1331 (1993).
16. J. Bicerano, E. R. Eidsmoe, and J. T. Seitz, SYNTHIA software available from Biosym Technologies, Inc., San Diego, California.
17. J. Bicerano, *Prediction of Polymer Properties*, Marcel Dekker, New York, 1993.
18. D. W. van Krevelen, *Properties of Polymers: Their Estimation and Correlation with Chemical Structure, (a) 2nd ed.*, Elsevier, Amsterdam, 1976; (b) 3rd ed., Elsevier, Amsterdam, 1990.
19. C. B. Arends, *J. Appl. Polym. Sci. 49*:1931 (1993).
20. L. C. E. Struik, *Physical Aging in Amorphous Polymers and Other Materials*, Elsevier Scientific Publishing, Amsterdam, 1978.

21. M. R. Tant and G. L. Wilkes, *Polym. Eng. Sci.*, *21*:874 (1981).
22. R. E. Robertson, *Computational Modeling of Polymers* (J. Bicerano, ed.), Marcel Dekker, New York, 1992, chapter 6.
23. N. Neki and P. H. Geil, *J. Macromol. Sci.—Phys. B8*:295 (1973).
24. R. A. Bubeck, S. E. Bales, and H.-D, Lee, *Polymer Engineering and Science*, *24*:1142 (1984).
25. T. G. Fox and S. Loshaek, *J. Polym. Sci. 15*:371 (1955).
26. S. Loshaek, *J. Polym. Sci. 15*:391 (1955).
27. H. Stutz, K.-H. Illers, and J. Mertes, *J. Polym. Sci., Polym. Phys. Ed. 28*:1483 (1990).
28. C. G. Reid and A. R. Greenberg, *ACS Div. Polym. Mater. Sci. Eng. Preprints, 56*:764 (1987).
29. G. C. Martin and M. Shen, *ACS Polym. Preprints, 20*:786 (1979).
30. R. A. Pearson and A. F. Yee, *J. Mater. Sci. 24*:2571 (1989).
31. V. Bellenger, B. Mortaigne, and J. Verdu, *J. Appl. Polym. Sci. 44*:653 (1992).
32. U. T. Kreibich and H. Batzer, *Angew. Chem. 83*:57 (1979).
33. B. Fuller, J. T. Gotro, and G. C. Martin, *Adv. Chem. 227*:215 (1990).
34. L. R. G. Treloar, *The Physics of Rubber Elasticity, 3rd ed.*, Clarendon Press, Oxford, 1975, chapter 8.
35. W. W. Graessley and S. F. Edwards, *Polymer 22*:1329 (1981).
36. S. Wu, *Polym. Eng. Sci. 30*:753 (1990).
37. A. E. Woodward, *Atlas of Polymer Morphology*, Hanser Publishers, Munich, 1989.
38. S. W. Tsai and H. T. Hahn, *Introduction to Composite Materials*, Technomic, Lancaster, PA, 1980, chapter 9.
39. S. W. Tsai, *Composites Design—1985*. Think Composites, Dayton, OH, 1985.
40. T. S. Chow, *J. Mater. Sci. 15*:1873 (1980).
41. C. B. Arends, *Polym. Eng. Sci. 32*:841 (1992).
42. J. Bicerano and S. R. Ovshinsky, *Applied Quantum Chemistry* (V. H. Smith, H. F. Schaefer III, and K. Morokuma, eds.), D. Reidel, Holland, 1986, pp. 325–345.
43. J. Bicerano and D. Adler, *Pure Appl. Chem. 59*:101 (1987).
44. M. F. Thorpe, *J. Non-Cryst. Solids 57*:355 (1983).
45. M. Takayanagi, *Molecular Basis of Transitions and Relaxa-*

tions (D. J. Meier, ed.), Gordon & Breach, New York, 1978, pp. 117–165.

46. R. J. Samuels, *Chemtech*, March 1974, pp. 169–177, and references therein.
47. J. T. Seitz, *Encyclopedia of Polymer Science and Engineering*, *16*:154 (1989).
48. S. W. Tsai and N. J. Pagano, *Composite Materials Workshop*, Technomic, New York, 1968, pp. 233–253.
49. S. G. Advani and C. L. Tucker III, *J. Rheol. 31*:751 (1987).
50. P. I. Vincent, *Polymer 13*:558 (1972).
51. E. H. Andrews and L. Bevan, *Polymer 13*:337 (1972).
52. E. H. Andrews, G. M. Levy, and J. Willis, *J. Mater. Sci. 8*:1000 (1973).
53. E. J. Kramer, *Adv. Polym. Sci. 52/53*:1 (1983).
54. E. J. Kramer and L. L. Berger, *Adv. Polym. Sci. 91/92*:1 (1990).
55. R. P. Kambour, *Polym. Commun. 24*:292 (1983).
56. M. T. Takemori, R. P. Kambour, and D. S. Matsumoto, *Polym. Commun. 24*:297 (1983).
57. G. H. Michler, *J. Mater. Sci. 25*:2321 (1990).
58. A. A. Griffith, *Phil. Trans. Roy. Soc. A221*:163 (1920).
59. N. Brown, *J. Mater. Sci. 18*:2241 (1983).
60. M. A. Maxwell and A. F. Yee, *Polym. Eng. Sci. 21*:205 (1981).
61. A. Margolina and S. Wu, *Polymer 29*:2170 (1988).
62. J. D. Ferry, *Viscoelastic Properties of Polymers*, John Wiley, New York, 1961.
63. C. J. G. Plummer and A. M. Donald, *J. Polym. Sci., Polym. Phys. Ed. 27*:325 (1989).
64. T. C. B. McLeish, C. J. G. Plummer, and A. M. Donald, *Polymer 30*:1651 (1989).
65. B. Rosen, ed., *Fracture Processes in Polymeric Solids*, Interscience, New York, 1964.
66. O. M. Ettouney and C. C. Hsiao, *J. Appl. Phys. 64*:4884 (1988).
67. C. L. Dym, *Comput. Struct. 16*:101 (1983).
68. T. H. H. Pian, *RCA Rev. 39*:648 (1978).
69. J. Bicerano, *Prediction of Polymer Properties*, Marcel Dekker, New York, 1993, Section 11.C.4.
70. C. Bauwens-Crowet, J.-C. Bauwens, and G. Homes, *J. Mater. Sci. 7*:176 (1972).

71. J. Haussy, J. P. Cavrot, B. Escaig, and J. M. Lefebvre, *J. Polym. Sci. Polym. Phys.* Ed. *18*:311 (1980).

72. P. Beardmore, *Philos. Mag 19*:389 (1969).

73. A. Thierry, R. J. Oxborough, and P. B. Bowden, *Philos. Mag. 30*:527 (1974).

74. C. Bauwens-Crowet, *J. Mater. Sci. 8*:968 (1973).

75. P. I. Vincent, *Plastics 27(August)*:105 (1962).

76. R. E. Robertson, *Appl. Polym. Symp. 7*:201 (1968).

77. P. B. Bowden and S. Raha, *Philos. Mag. 22*:463 (1970).

78. R. D. Andrews and W. Whitney, Report No. TD-123-64, Textile Division, Department of Mechanical Engineering, Massachussetts Institute of Technology, Cambridge, 1964.

79. Y. Imai and N. Brown, *Polymer 18*:298 (1977).

80. A. S. Argon and M. I. Bessonov, *Philos. Mag. 35*:917 (1977).

81. S. Yamini and R. J. Young, *J. Mater. Sci. 15*:1814 (1980).

82. C. Bauwens-Crowet, J.-C. Bauwens, and G. Homes, *J. Polym. Sci. A-2,7*:735 (1969).

83. W. Wu and A. P. L. Turner, *J. Polym. Sci., Polym. Phys. Ed. 13*:19 (1975).

84. J. T. Ryan, *Polym. Eng. Sci. 18*:264 (1978).

85. S. S. Sternstein, L. Ongchin, and A. Silverman, *Appl. Polym. Symp. 7*:175 (1968).

86. J. A. Roetling, *Polymer 6*:311 (1965).

87. J. A. Roetling, *Polymer 6*:615 (1965).

88. R. N. Haward, B. M. Murphy, and E. F. T. White, *J. Polym. Sci. A-2, 9*:801 (1971).

89. J. Bicerano, *Prediction of Polymer Properties*. Marcel Dekker, New York, 1993, references for chapter 11.

90. N. Brown, *Principles of Polymer Engineering* (N. G. McCrum, C. P. Buckley, and C. B. Bucknall, eds.), Oxford University Press, New York, 1988, chapter 6.

91. A. S. Krausz and H. Eyring, *Deformation Kinetics*, John Wiley, New York, 1975.

92. J. C. M. Li, C. A. Pampillo, and L. A. Davis, *Deformation and Fracture of High Polymers* (H. H. Kausch, J. A. Hassell, and R. I. Jaffee, eds.), Plenum Press, New York, 1974, pp. 239–258.

93. Y. Termonia and P. Smith, *Macromolecules 20*:835 (1987).

94. Y. Termonia and P. Smith, *Macromolecules 21*:2184 (1988).

95. J. L. Amos, L. C. Rubens, and H. G. Hornbacher, *Styrene: Its Polymers, Copolymers and Derivatives* (R. H. Boundy and R. F. Boyer, eds.), Reinhold, New York, 1952, pp. 723–729.

96. R. A. Pearson and A. F. Yee, *J. Mater. Sci.* 24:2571 (1989).

97. D. J. Meier, *Thermoplastic Elastomers: A Comprehensive Review* (N. R. Legge, G. Holden, and H. E. Schroeder, eds.), Hanser Publishers, Munich, 1987, chapter 11.

98. R. J. Samuels, *J. Macromol. Sci. Phys. B4*:701 (1970).

99. R. J. Samuels, *J.. Polym. Sci. A2, 10*:781 (1972).

100. R. J. Samuels, *J. Macromol. Sci. Phys. B8*:41 (1973).

101. L. J. Gibson and M. F. Ashby, *Cellular Solids*, Pergamon Press, New York, 1988.

102. (a) W. C. Kuryla, F. E. Critchfield, L. W. Platt, and P. Stamberger, *J. Cellular Plast.* 2(2):84 (1966); (b) M. A. Koshute and H. A. Freitag, *Second Generation PHD Polyol for Automotive Flexible Molding*, Proceedings of the SPI/FSK Polyurethanes World Congress, Aachen, West Germany, September 1987, Technomic Press, Lancaster, PA.

103. W. W. Adams, R. K. Eby, and D. E. McLemore, eds., *The Materials Science and Engineering of Rigid-Rod Polymers*, Materials Research Society, Pittsburgh, 1989.

104. J. Bicerano, ed., *Computational Modeling of Polymers*, Marcel Dekker, New York, 1992.

105. R. J. Roe, ed., *Computer Simulation of Polymers*, Prentice-Hall, Englewood Cliffs, NJ, 1991.

106. C. A. Coulson, *Valence*, 2nd ed., Oxford University Press, New York, 1961.

107. L. Pauling, *The Nature of the Chemical Bond*, 3rd ed., Cornell University Press, Ithaca, NY, 1973.

108. Science and Engineering on Cray Supercomputers: Proceedings of the Third International Symposium, Cray Research, Minneapolis, 1987.

2
Multiphase Toughening of Plastics

CHARLES B. ARENDS* The Dow Chemical Company, Midland, Michigan

I. INTRODUCTION

Having multiple phases in materials is certainly not new. Such structures have existed as long as there has been advanced life on Earth. Palm leaves, bones, and beetle exoskeletons are all examples of materials that depend on multiphase structure for their particular combination of properties. Then Man gets into the act. He begins with Damascus swords and continues today with a range of multiphase materials, from structural composites to foams (like semiconductor theorists, I include holes as a component of multiphase structures).

Let us define a multiphase polymer system as one in which there are two or more demonstrably different regions or sets of regions in the material under study. The material may be chemically homogeneous as in semicrystalline polymers, it may consist of chemically bound segments of differ-

*Retired

ent monomers as in block copolymers, it may be an assembly of different components as in fiber-reinforced composites or rubber-modified thermoplastic glasses, or it may be a higher-order combination of these materials as in fiber-reinforced semicrystalline polymers.

Multiphase materials are usually developed in order to give *combinations* of properties that are superior to single component materials. These properties can include toughness, heat distortion, processability, cost, or any of a host of other properties we deem necessary. For example, as will be shown later, it frequently is a straightforward matter to increase the toughness of a polymer by decreasing its glass transition temperature. But if we are able to toughen our material by using a dispersed toughening phase, we can often achieve the combination of heat resistance and toughness we are looking for without changing base polymers. In practice, we usually have to balance the advantages gained against some disadvantages such as modulus reduction or loss of optical clarity, but a useful balance of properties can frequently be found.

The purpose of this chapter is to present some unifying thoughts on the "why's" and "how's" of multiphase systems. We will examine specific techniques for utilizing discrete phases. We will consider the factors that go into choosing a technique and give some examples of the technique. For the time being we will not consider any interphase that may exist between different components except as it contributes directly to energy absorption.

II. MECHANISMS

Three basic mechanisms of energy absorption are available within a single material: shear yielding, crazing, and crack formation. The deformation associated with the particular mechanism occurs irreversibly on the time scale of the deformation process. It is entirely possible that crazing or shear yielding can be reversed by heating the deformed sample to the glass transition or crystalline melting temperature of the polymer, but on the time scales of most mechanical processes the deformation is essentially fixed. It is even possible to reestablish continuity across a crack boundary [1], but the procedure is much more involved.

Shear yielding is the irreversible change of shape of the material under stress. It may occur over the entire stressed region of material or it may occur in localized portions of the stressed region, in which case we frequently refer to the response as shear "banding." An example of shear banding is shown in Fig. 1.

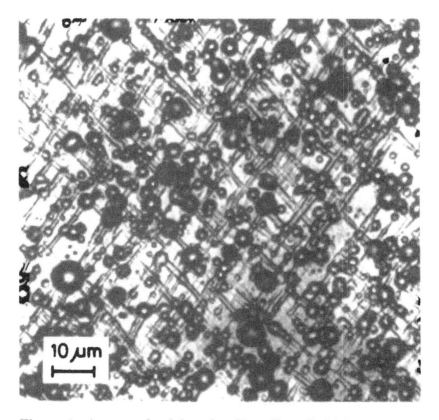

Figure 1   An example of shear banding. (From Ref. 7.)

Crazing is a localized irreversible volume expansion. It is often seen in tensile experiments as optical discontinuities that resemble cracks. The craze is not a crack however, as can be shown by microscopic examination. Figure 2 reveals the cross section of a craze showing its discontinuous structure and expanded volume. We will discuss crazing in more detail later. Crack formation is the total separation of matter along a surface. The surface may encompass the entire cross-sectional area of a specimen or it may form locally without causing complete failure.

All three of the preceding mechanisms may be available to a material under different circumstances. Figure 3 shows a crazing envelope superimposed on a yield envelope for poly(methylmethacrylate) (PMMA) [2]. This composite graph

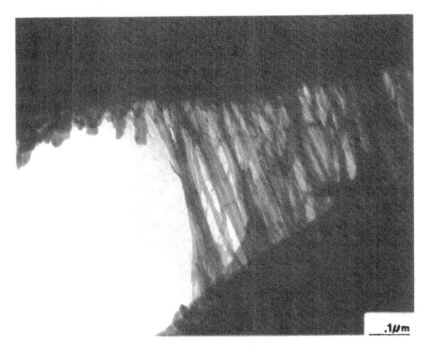

Figure 2   An example of craze fibrils. (From Ref. 21.)

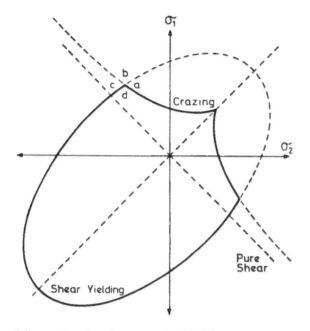

Figure 3 Crazing cusp for PMMA. (From Ref. 2.)

shows that under most net dilative [1] stresses PMMA will craze and under net compressive stresses it will shear yield. At sufficiently low temperatures neither shear yielding nor crazing will occur and only crack formation will be available [3]. Of the three, shear yielding and crazing offer much higher energy absorption potential and are consequently to be preferred, if they can be brought into play. The higher potential arises because large numbers of shear bands or crazes can be activated under load, while only a few crack planes are normally found at failure.

We have several other mechanisms to consider in which we may actually alter the mode of material response by mechanical means. Crazing may be sacrificed to induce shear yielding. Notch sensitivity can be significantly reduced. The effective size scale can be changed so that we can operate in plane stress instead of plane strain. These too will be discussed in more detail later in this chapter.

III. CHOOSING A POLYMER

When we employ multiphase structures, we either try to use a separate phase to nucleate the energy absorption processes of another (as in rubber-modified styrenics) or we attempt to combine the desirable properties of each of the components to give a superior combination of properties (as in interpenetrating networks or lamellar structures). The approach taken will depend on many factors that may or may not have anything to do with toughness. In most cases we do not have complete freedom to choose which polymer will be used as a base material, but rather we are often limited by economic or additional performance constraints. As a result we are forced to choose from a restricted list of materials and the best we can hope to do is to optimize our chances for success. We do this by determining which member(s) on that list will be most amenable to toughening using the mechanisms available.

It turns out that we can usually do a more effective job of "toughening" a polymer, regardless of the mechanism, when the base polymer has some intrinsic toughness. We can assess this potential by measurements on the unmodified resin. If the material exhibits a tensile yield in the unmodified form, we have about all we can hope for in the way of intrinsic toughness. Our job then is to see that we do not embrittle the system when we add separate phases for other reasons. (See discussion of plane stress in Section VI.)

For very brittle materials we typically use some method of measuring fracture energy (G1c) to evaluate toughness. One such determination is demonstrated in Fig. 4 using a double-cantilever-beam specimen. In these measurements we look at the amount of energy used by the specimen in forming the crack surface. The higher the energy, the more toughness

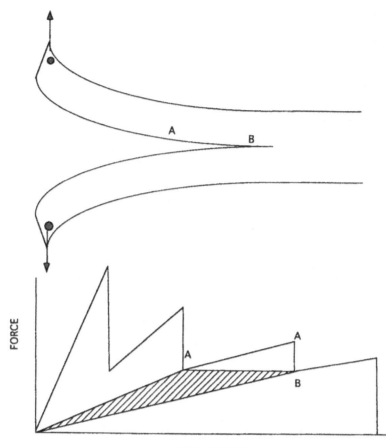

FORCE

DISPLACEMENT

Figure 4 Double-cantilever method of determining fracture energy. The shaded area is a measure of the work needed to propagate a crack from A to B at high speed (the arrest energy).

we should be able to build into our system. One straightforward way to think about this is to realize that there is usually some material distortion in the vicinity of the growing crack [4]. The higher the fracture energy, the more distortion is generated and the more potential there is for further toughening. We will not go into the details of how the experiments are run or how they are analyzed, since the field of fracture mechanics is outside the scope of this chapter. If the reader wishes to study further, there are a number of excellent texts from which to choose [5].

An alternative measurement produces a parameter I choose to call "ductility potential." This measurement [6] is particularly useful when the polymer has very high fracture energies. In this procedure we measure the compressive yield strength of the material with tensile loads applied perpendicular to the compression. When a standard testing machine is used, the yield will be defined by the shape of the (compressive) stress–strain curve. In local practice a constant tensile load is applied to a thin strip of polymer, and the rate of creep through the thickness is monitored under increasing compressive load until a rate of 0.0005 sec^{-1} is reached. This is arbitrarily defined as the biaxial yield condition, although we frequently observe creep rate acceleration around this value that would certainly indicate yield. The maximum tensile load at which the yield point is reached without sample failure is divided by the (extrapolated) tensile yield strength to give the "ductility potential" (Fig. 5). Once again, the higher the number, the more effectively we can toughen the material.

For processes that involve crazing, the determination is not quite so clear. For example, polystyrene has a higher fracture energy than styrene-co-acrylonitrile (SAN), but a lower ductility potential. In this case, ductility potential wins as the better indicator of toughenability, and for reasons that will be given later, rubber-modified SAN (acrylonitrile-butadiene-styrene, ABS) will typically have a higher impact resistance than rubber-modified polystyrene (high-impact polystyrene, or HIPS). In ABS, we can actually invoke both crazing and shear yielding either separately or simultaneously depending on the details of the dispersed-phase morphology and the test conditions [7].

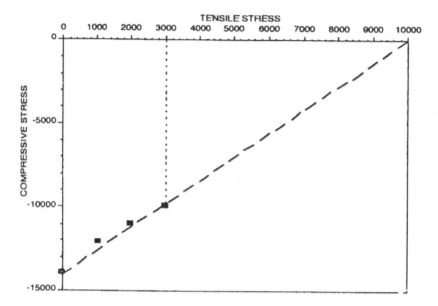

Figure 5   Evaluation of "ductility potential."

IV.   IMPLEMENTING THE MECHANISMS

A.   Crazing

Crazing is a highly successful mechanism for toughening some polymers especially for vinyl polymers. In order for the mechanism to be successful, however, a very large number of crazes must be generated. Referring to Fig. 2, we see that a craze consists of highly drawn threads of polymer associated with empty space to a net density of approximately 0.5 [8]. This means that the craze occupies roughly double the volume of the undrawn polymer and it achieves this by doubling only one dimension. The result is that we have energy absorption via a thread-drawing process. We can evaluate the energy absorbed per craze according to:

$$e = s_y D_l A_c$$

where e is the absorbed energy per craze, s_y is the yield strength. D_l is the change in local sample length per craze, and A_c is the area per craze. Using representative values for the input parameters of 7×10^7 Pa, 10^{-6} m, and 10^{-7} m^2 respectively yields about 10^{-5} J/craze. For impact polystyrene we calculate absorbed energies under slow-speed conditions of about 10^5 J/cm^3. This amounts to 10^{10} crazes/cm^3. In unmodified polystyrene we typically observe about 100 crazes/cm^3 during a tensile experiment. Thus, the few crazes there cannot contribute significantly to energy absorption. This is why we turn to rubber modification.

The addition of small, well-dispersed rubber particles can increase the number of crazes generated by many orders of magnitude. To demonstrate the process, a typical impact polystyrene plastic will have about 30% gel at a 4-μm average rubber particle diameter, from which we calculate 10^{10} particles/cm^3. Note that these numbers do not refer to any particular HIPS material, but are merely intended to be representative. This simple demonstration shows that the orders of magnitude are similar and are appropriate for a material in which the rubber phase acts as a craze multiplier. The mechanism of initiation is simply for the rubber particles to act as a stress concentrator. As stress concentrators, the rubber particles cause crazes to nucleate in the elevated stress zones in the vicinity of the particles (Fig. 6) [9]. The crazes then grow into lower-stress regions and stabilize, thus permitting many crazes to be generated.

A typical ABS will utilize smaller particles than HIPS by factors of 4 to 10. Consequently, up to 1000 times the number of particles may be available. Even though the fracture energy for polystyrene can be six times that of SAN [10], the superior number of particles in ABS more than makes up for it. The choice of particle size is based on the sizes that maximize performance and is different for each base polymer. Test

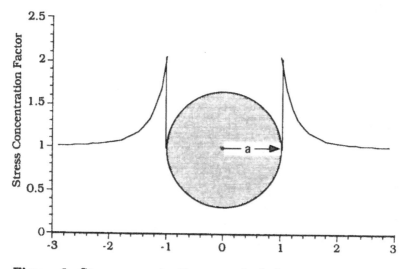

Figure 6 Stress concentration around a hole.

rate, the incorporation of plasticizers, interface effectiveness, average particle size, as well as size distribution all contribute to the efficiency of craze generation and must be considered when designing a craze-based impact-resistant plastic. For further details see Chapter 8.

B. Crack Multiplication

Just as we employ a fine dispersion of rubber particles to nucleate crazes, we can likewise use rubber particles to nucleate crack initiation sites. Figure 7 shows multiple fracture planes arising from the addition of rubber particles to brittle thermoset castings. The rubber particles once again provide stress concentration sites above and below the growing crack plane and form local cracks that in turn absorb energy. Even though the number of planes that can be activated is small by comparison with the number of crazes that can be generated in a thermoplastic, this technique is still viable for improving

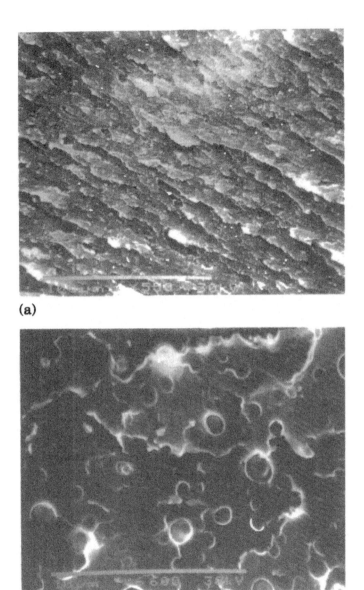

(a)

(b)

Figure 7 Fracture surfaces of rubber-modified thermosets. (a) DER™ 332 cured with methylene dianiline. (b) DEN™ 438 cured with diamino diphenyl sulfone. (™ Trademark of The Dow Chemical Company.)

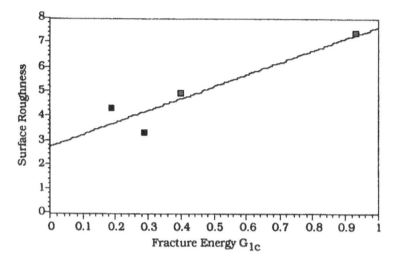

Figure 8 Surface roughness versus fracture energy for several cured epoxies.

toughness. As an illustration of the technique a series of rubber-modified thermosets were prepared and fractured. The surface roughness of each was measured and compared with fracture energy (G_{1C}). The results are plotted in Fig. 8 and indicate a definite correlation between the two. Incidentally, fracture energies in the rubber-modified samples were approximately 2.5 times the value for a neat sample of the matrix. In other rubber-modified thermosetting systems, fracture energies of from 1.1 to 7 times the neat matrix energies have been observed. Higher toughening efficiencies are usually observed for intrinsically tougher matrices.

C. Shear Banding

Referring once again to Fig. 1, we see a host of shear bands developed in conjunction with rubber particles. As before, the rubber particles provide stress field perturbations, but the rubber by itself adds an extra feature, compression of the matrix.

In order to understand this effect we need to realize that the thermal expansion coefficients of glassy plastics are only about one-third that of rubbery plastics. Therefore, when the multiphase material is cooled after processing, a thermal expansion mismatch occurs below the glass transition temperature of the matrix. This causes the rubber phase to be in hydrostatic tension and the matrix to be in compression (Fig. 9). This in turn biases the material response toward shear banding and away from crazing. Matsushiga et al. [11] have demonstrated the validity of the concept by applying external pressure to polymers and observing the change in energy to break under tensile loading.

Referring to Fig. 3 we see that the addition of a compressive component will move a simple tensile stress field off the x axis into the (biaxial) second or fourth quadrant. If the compressive stress is high enough, response can be shifted from

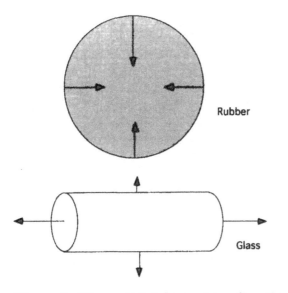

Figure 9 Thermal shrinkage stress directions developed by differential contraction in rubber and glass during cooling.

crazing to shear yielding. For example, in unmodified polystyrene the ratio of tensile stresses needed to generate shear banding to the stresses needed to cause crazing has been calculated to be 1.2. Compression resulting from thermal expansion mismatch can reduce the tensile stresses needed to generate shearing while increasing the stresses needed to bring about crazing. It has been predicted that the transition from crazing to shear banding in rubber-modified polystyrene will occur (at slow strain rates) when the rubber content reaches 25%. This agrees well with the occurrence of accelerated increase in impact resistance as rubber content increases above 25%.

When considering the competition between crazing and shearing, it is important to remember that the yield envelope for a given polymer is temperature and rate dependent. It is actually possible to have crazing at high strain rates and shear banding at low strain rates in the same plastic [7].

V. MORE ABOUT THERMAL STRESSES

It is extremely important that thermal stress concepts be understood since embrittlement can also result from thermal expansion mismatch. If, instead of adding rubber to a brittle polymer, we add chopped glass to a ductile polymer, a loss of impact resistance is possible. In the case of polycarbonate, the addition of chopped glass causes the impact resistance to drop to one-eighth of its neat value [12]. The effect of adding glass to a polymer is the opposite of adding rubber since glass has an extremely low thermal expansion. The net result is to bring the polymer into a brittle state due to expansion rather than a ductile one via compression. Consequently, one would expect that the addition of rubber to the polymer–glass mixture would restore toughness. Figure 10 shows the results. In this experiment only enough rubber was added to balance the thermal mismatch of the glass to the polycarbonate. Even though only a partial recovery of impact resistance was

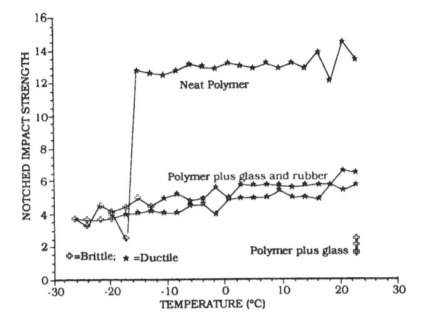

Figure 10 Effect of chopped glass and rubber latex on impact strength in polycarbonate.

achieved, the failure mode was restored to ductile by the addition of rubber. To fully appreciate the effects of thermal stress we need to understand the concepts of plane stress and plane strain, which is discussed next.

VI. OTHER APPROACHES

By definition plane stress is a stress condition in which there is stress in only a single direction [13]. Conversely, plane strain refers to a stress condition in which stresses naturally develop perpendicular to an applied stress. This results in a triaxial stress state toward the center of the specimen that opposes shear flow and results in embrittlement of the speci-

men. Typically one observes plane stress in very thin specimens and plane strain in thick. Figure 11 is a pictorial representation of how plane strain can arise under uniaxial load
[14]. From this picture one sees a shear response near the
sample surface and a brittle response in the center. In a practical experiment the low energy state dominates. The shear response dimension is material dependent (intrinsic), *not* sample
dependent (extrinsic); that is, it does not depend on the size of
the piece being tested. Thus, if the sample is thin enough, it
only experiences the plane stress response near the surface. As
a result, we have relatively high energy absorption potential
in thin sections and relatively low potential in thick. The absolute magnitude of the plane stress dimension varies greatly
from polymer to polymer, from millimeters in polycarbonate to
submicrons in polystyrene. Another technique for utilizing
plane strain to plane stress conversion has to do with using a

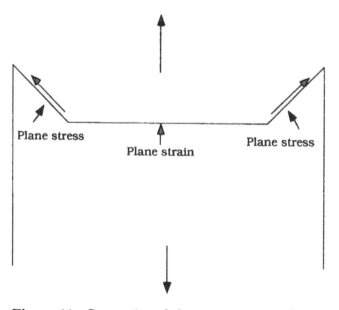

Figure 11   Generation of plane strain via surface traction.

second phase to reduce the matrix dimensions internally. Wu [15] has demonstrated the effect of interparticle distance on the toughness of rubber-modified nylon. He observed a critical maximum spacing for toughness in the mixture and relates this observation to plane strain to plane stress conversion. In keeping with the concept of a critical dimension for plane stress we can readily see that the formation of thin ligaments between second-phase particles can provide a significant contribution to energy absorption potential. Yee et al. [16] have demonstrated the process in polyethylene-modified polycarbonate. This mechanism seems to be most useful for materials that are normally ductile but can be embrittled easily.

One does not normally consider trying to convert a material response from plane strain to plane stress, but this is part of the explanation of the effect of thermal stress on rubber-modified polymers. Darwish et al. [17] have shown that the dimension which allows the transition can be qualitatively related to a ductile zone size given by

$$r_y = \frac{(K_{1C}/s_y)^2}{2P}$$

where r_y is the zone size, K_{1C} is the stress intensity factor, which is related to the fracture energy, and s_y is the yield stress. As a general rule, the addition of a rubber to a brittle matrix increases K_{1C} and decreases s_y so that the ductile zone size is increased and the plane stress condition is more likely.

VII. REDUCTION OF NOTCH SENSITIVITY

In addition to identifying plane strain to plane stress transition, Yee et al. [16] commented that polyethylene-modified polycarbonate exhibited low notch sensitivity. Notch sensitivity is simply the effect that a mechanical imperfection (notch) has on the properties of a material. Typically, the introduction of a notch reduces the mechanical strength or "toughness" depending on the ratio of the depth of an imperfection to the radius of curvature at the tip of the notch; that is,

$$s = s_0 \left(1 + 2 \sqrt{\frac{a}{r_0}} \right)$$

where s is the maximum stress at the tip of the crack, s_0 is the far-field stress, a is the notch depth, and r_0 is the radius of curvature at the tip of the notch. Therefore, a smaller radius or a greater depth should result in an increased stress concentration and a loss in the macroscopic properties. A typical set of data is shown in Fig. 12 [18] where the impact strength of poly(vinyl chloride) is seen to increase with notch radius. However, we see that there is virtually no notch sensitivity for two

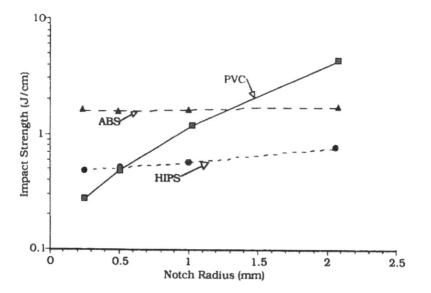

Figure 12 Effect of rubber modification on notch sensitivity. (After Ref. 18.)

rubber-modified styrenic polymers. One way to think of the effect of the second-phase rubber particles is to consider that the rubber is typically dispersed on a scale that is approximately two orders of magnitude smaller than the notch radius. Consequently, stress in the vicinity of the notch is controlled by yield phenomena that are initiated by the particles and cannot grow to bring about an unstable state as a result of the notch. Alternatively, we could say that soft particles convert plane strain to plane stress by defining thin ligaments between the particles [15].

VIII. CONTINUOUS TOUGH LAYERS

Special structures can be produced via clever polymerization or processing, in which an intrinsically tough material forms a continuous layer or network. Other components may or may not be continuous as well. Figures 13 and 14 show examples of these genre. In Fig. 13 we see a lamellar structure that was produced by multilayer coextrusion [19]. Both materials are continuous in parallel planes. This structure has shown an unusual ability to produce tough materials when one component is intrinsically tough, the other is brittle, and the layers are sufficiently thin [20]. Figure 14 shows a thermoplastic-toughened thermoset. Here the thermoplastic is continuous and the thermoset dispersed even though the fraction of thermoplastic is less than 0.25. Perhaps the best way to describe this structure is as a highly filled thermoplastic, and its properties are representative of this type of structure.

IX. SUMMARY

Polymer toughness is a complicated concept. Toughening a resin by the addition of additional phases is even more complicated. The additional phase must be added in a form that

Figure 13   Lamellar structure from coextrusion (From Ref. 19.)

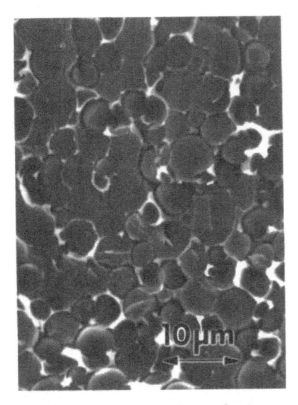

Figure 14 Continuous thermoplastic network with dispersed thermoset.

makes use of the inherent toughness in the base polymer. It can be added as a discrete phase as in impact polystyrene or ABS. It can be incorporated as a co-continuous phase as in interpenetrating networks (IPN).

Dispersions of toughening agents are used to nucleate whatever energy-absorbing mechanism is available. Frequently, this will be crazing or shear banding in thermoplastics and multiple crack formation in thermosets. Other mechanisms are possible, especially in thermosets. Co-continuous structures will often use the properties of the second phase to balance those of the first, thereby resulting in a plastic with superior overall performance when compared with either component.

REFERENCES

1. R. P. Wool, B-L. Yuan, and O. J. McGarel, *Polym. Eng. Sci.* 29:1340 (1989).
2. S. Sternstein and L. Ongchin, *Polym. Preprints 10*(No. 2) (1969).
3. P. I. Vincent, *Polymer 13*:558 (1972).
4. A. Moet, *Fracture of Plastics* (W. Brostow and R. D. Corneliussen, Eds.), Hanser Publishers, New York, 1986, chapter 18.
5. (a) A. J. Kinloch and R. J. Young, *Fracture Behavior of Polymers*, Elsevier Applied Science Publishers, New York, 1985; (b) M. F. Kaaninen and C. H. Popelar, *Advanced Fracture Mechanics*, Oxford University Press, New York, 1985.
6. P. B. Bowden and J. A. Jukes, *J. Mater. Sci. 3*:183 (1968).
7. C. B. Bucknall, *Toughened Plastics*, Applied Science, London, 1977.
8. R. P. Kambour, *J. Polym. Sci. A2*:4159 (1964).
9. S. Timoshenko and J. N. Goodier, *Theory of Elasticity*, McGraw-Hill, New York, 1951, p. 361.
10 C. B. Arends, *J. Appl. Polym. Sci. 10*:787 (1966).
11. K. Matsushiga, S. V. Radcliffe, and E. Baer, *J. Appl. Polym. Sci. 30*:1853 (1976).
12. The Dow Chemical Company, Form No. 301-643-485, Midland, MI.
13. S. Timoshenko, and J. N. Goodier, *Theory of Elasticity*, McGraw-Hill, New York, 1951, chapter 2.
14. A. S. Tetelman and A. J. McEvily, Jr., *Fracture of Structural Materials*, John Wiley, New York, 1967.
15. S. Wu, *Polymer 26*:1855 (1985).
16. A. F. Yee, W. V. Olszewski, and S. Miller, *Toughness and Brittleness of Plastics* (R. D. Deanin and A. M. Crugnola, eds.), Advances in Chemistry Series No. 154, American Chemical Society, Washington, DC, 1976.

17. A. Y. Darwish, F. J. Mandell, and F. J. McGarry, Research Report R82-2 (1982), Department of Materials Science and Engineering, Massachusetts Institute of Technology, Cambridge, 1982.
18. A. C. Morris, *Plast. Polym.* 36:433 (1968).
19. W. J. Schrenk, R. A. Lewis, J. A. Wheatley, and C. B. Arends, 7th Annual Meeting, Polymer Processing Society, Hamilton, Ontario, Canada, 1991.
20. A. Hiltner, K. Sung, E. Shin, S. Bazhenov, J. Im, and E. Baer, *Mater. Res. Soc. Symp. Proc.* 255:141 (1992).
21. M. J. Yaffee and E. J. Kramer, *J. Mater. Sci.* 16:2130 (1981).

3
GRAFTED RUBBER CONCENTRATES

DALE M. PICKELMAN* The Dow Chemical
Company, Midland, Michigan

I. INTRODUCTION

Synthetic polymers that are brittle at or below room temperature need additives to expand their toughness over a broad temperature range. Some early examples of such brittle polymers are glyptal (glycerol–phthalic anhydride) resins, polyvinyl chloride (PVC), and polystyrene (PS). Patton [1] listed some commonly used polymer additives with alkyds (glyptal + fatty oil) for improvement in toughness. These polymer additives are nitrocellulose, urea-formaldehyde, melamine-formaldehyde, chlorinated rubber, and chlorinated parafins. Other additives were incorporated into the resin system by way of chemical modification, such as styrene, phenolics, silicones, epoxies, and isocyanates. Desirable property improvements for these modified alkyd resins, in addition to impact resistance,

---

*Retired

include chemical resistance (alkali and water), hardness, color, weather resistance, and durability. PVC and PS are higher molecular weight (MW) brittle molding polymers and also need physical or chemical modification to improve medium- and low-temperature impact strength, with little change in dielectric properties and with long-term durability over cyclical stress conditions. New product families have been generated by rubber toughening of high-performance thermoplastic polymers, such as aromatic polyesters, polycarbonates, nylons, and styrene-acrylonitrile (SAN) copolymers. High-performance thermoset polymers, such as polycyanate esters and epoxy-modified resin hardner systems, can also be rubber toughened to broaden the window of temperature performance and durability. Developing understanding of the mechanisms of toughening and failure will enable one to predict product durability or product requirements for new applications.

The following outline of polymers and toughening agents illustrates the many and various systems currently involved in toughening thermoplastics and thermosets. Seymour [2] describes the systems indicated in Tables 1 to 12 in detail and includes numerous references to them.

As noted, a simple and early toughening method was to blend a low-MW modifier with the brittle polymer [1,2]. This works to a limited extent, but was soon replaced with either random copolymers or polymer blends with copolymers. The next improvements were to chemically modify the rubber in situ during polymer synthesis or in a separate reaction prior

Table 1 Low Molecular Weight Additives for Toughening Brittle Condensation Polymers

Polymer	LMW additive
Cellulose nitrate	Camphor
Casein formaldehyde	Moisture
Glycerol phthalate resin	Unsaturated oils

Table 2 Additives for Rubber Toughening of Brittle Vinyl Polymers

Polymer	Rubber
Poly(vinylchloride), PVC	Nitrile-butadiene-rubber, NBR
Poly(styrene), PS	Styrene-butadiene-rubber, SBR
Poly(styrene-acrylonitrile)SAN	NBR
Poly(styrene–maleic anhydride), SMA	NBR

Table 3 High-Performance Polymers

Acetals
Acrylics
Nylons
Polyesters
Polyolefins
Thermosets (epoxy, novolac, cyanate)

Table 4 Examples of Polyvinyl Chloride Systems

PVC resin + plasticizers (LMW)
Copolymers:
   Vinyl chloride (VC)/vinyl acetate (VA)
   VC/vinyl butyl ether (VBE)
   VC/vinylidene chloride (VDC)
PVC resin + copolymers:
   PVC + B/DEF (butadiene-diethylfumarate)
   PVC + B/MIPK (butadiene–methylisopropenyl ketone)
   PVC (latex) + B + AN (polymer in presence of latex)
   PVC + HDPE (high-density polyethylene)
PVC resin + graft copolymers:
   NBR (acrylonitrile-butadiene)
   ABS (acrylonitrile-butadiene-styrene)
   MBS (methyl methacrylate–butadiene–styrene)
   CPE (chlorinated polyethylene)
   Polyacrylates
   Polyurethanes
   EVA (ethylene-*g*-vinyl acetate)

Table 5 Examples of Modified Polystyrene Systems

S (styrene) + natural rubber	→	HIPS (high-impact PS)
S + SBR	→	HIPS
PS (latex) + SBR (latex)	→	HIPS

Table 6 Examples of Rubber-Modified Copolymeric Vinyl Thermoplastic Systems

SAN + NBR	→	ABS
S + AN (acrylonitrile) + PB (polybutadiene) (low AN, known as Dow's poor man's ABS)	→	ABS
SAN (latex) + NBR (latex) (better than solution properties)	→	ABS
SAN + SAN BR (graft core–shell rubber)	→	ABS
SAN + ASA (AN-acrylic-S)	→	ASA
SAN + ACS (AN–chlorinated polyethylene–S rubber)	→	ACS
S/MMA/AN + MBS	→	Clear ABS
S/αMS/AN + MBS	→	High temp. ABS
Polycarbonate + MBS + SAN	→	Modified ABS
S/AN/MA + NBR or ABS or ASA	→	Modified S/copolymer
SAN + OSA (olefin-S-AN)	→	Weatherable type ABS
SAN + ACS	→	Weatherable type ABS
SAN + ASA	→	Weatherable type ABS
PVC or PC (polycarbonate) + S-B-S or B-S-AN (block copolymers)		

Table 7 Styrene-Diene Block Copolymers Used as Modifiers

S-B pure diblock
SBS star or radial-branched
Linear SIS (styrene-isoprene-styrene)
Linear SBS (styrene-butadiene-styrene)
Weather resistant SBS + hydrogenated
Modified properties SBS + EPDM (ethylene-propylene-diene
   monomer)

Table 8 Core–Shell Acrylic Rubber Modifiers

MBS (core = SB, shell = PMMA) for
  PVC
  PC
  PMMA

Table 9 Polyolefin Rubber Modifiers

PIB (polyisobutylene)
IB/IP (isoprene) butyl-rubber
EPM (ethylene-propylene-copolymer)
EPDM (ethylene-propylene-diene monomer)
E + E-P (ethylene-propylene block copolymer)
Butyl-rubber-grafted polyethylene (ET polymer)

Table 10 High-Performance Condensation Polymers and Modifiers

Aliphatic polyesters
Aromatic polyester
  PET (polyethylene terephthalate)
  PBT (polybutylene terephthalate)
Polyester block copolymer (thermoplastic elastomer)
Acetals (polymers of formaldehyde)
Nylons
  Nylon 66 rubber-toughened ("supertough" nylon)
  Nylon 6 + EPDM

Table 11   Examples of Thermoset Systems and Modifiers

Novolac (phenol-formaldehyde) + polysulfide
Novolac + NBR
Epoxy (bisphenol A) + CTBN (carboxyl-terminated butadiene
  acrylonitrile)
Improved toughening (general procedures) for epoxies
  Excess amine-curing agent
  Chain extenders
  High MW amines
  Addition of bisphenol A

**Table 12   Summary of Polymer Systems: General-Purpose,
Engineering, Curable Polymers, Rubbers, and Impact Modifiers**

Thermoplastics	*Rubber Modifiers (continued)*
PS	Diene
SAN	Polybutadiene
SMA	SBR
PVC	S-B diblock
PVDC	S-I-S, S-B-S triblocks
Acrylics	S-B-caprolactone
Polyolefins	S-B-V pyridine
Polyesters	ABS (core–shell)
Acetals	NBR (core–shell)
Nylons	ASA (core–shell)
Polycarbonates	*Copolymer Modifiers*
Polyurethanes	Vinyl addition
Thermosets	VC/VA
Epoxy (bisphenol A)	VC/VDC
Novolac (phenol-formaldehyde)	VC/VBE
Cyanate (dicyclopentadiene ester)	E/VA
Rubber Modifiers	B/S/AN
Natural	B/DEF and B/MIBK
Butyl	Condensation
Polyethylene	PE diblock
HDPE	TPE
EPM, EPDM	
ET	
OSA	
ACS	

to the blending step. In both cases, graft rubber copolymers are formed and function as stabilizers for the rubber domains which allow dispersion throughout the continuous thermoplastic phase. The composition of the graft polymers attached to the rubber chains can differ from that in the continuous phase, but must be compatible for effective stabilization. One can readily envision many possible combinations and processes for rubber toughening.

As graft rubber technology developed, it became apparent that, in addition to grafted particles, a fraction (10 to 25% generally) of much larger rubber particles was needed. These large particles could not be economically produced by normal growth because of the relatively slow emulsion polymerization kinetics, except for the acrylate rubbers. Three routes have been pursued to attain these large particles for preparation of acrylonitrile-butadiene-styrene (ABS). These will be discussed in the order in which they were developed, but are briefly described as follows:

1.   One process involves agglomeration of small latex particles, early in the rubber polymerization process. This is accomplished by initially using a limited amount of water so that the monomer phase is the continuous one, with emulsion polymerization taking place in the suspended water droplets. This allows initial growth of particles with a minimum of soap since there is little shear in the droplets. Continual growth in the suspended water droplets causes a phase inversion (from oil to water continuous) that results in a controlled agglomeration. Addition of more soap stabilizes these newly formed particles and allows polymerization and subsequent grafting to occur with minimum formation of coagulum.

2.   Another process involves controlled or partial agglomeration of the latex at the end of the rubber polymerization and just prior to the grafting reaction. This process inherently forms a moderate amount of coagulum.

3.   A third process for producing large particles utilizes agglomeration when the graft rubber concentrate (GRC) is

first melt blended (usually with SAN resin in a compounding extruder). This agglomeration is controlled by the addition of nongrafted rubber latex or very lightly grafted rubber latex to normally grafted, monodisperse ABS latex. This technique has the added advantage of using coagulum-free latexes. The gloss level reached with this process is a few percent lower than that attained from the second method outlined above.

Overall, the core–shell rubber modifier systems have the versatility and characteristics, both physical and chemical, for toughening many polymers, both thermoplastic and thermoset systems.

II. TOUGHENING TECHNOLOGY AND GRC DEVELOPMENT

For the remainder of this chapter we will follow more or less chronological developments within broad major topics, beginning with the polymerization processes that produce the GRC particles.

A. Polymerization

Childers and Fisk [3] were among the first to be granted an early U.S. patent on the emulsion polymerization of ABS rubber modified copolymers.

Bradford and Vanderhoff [4] produced a GR-S latex with low gel and high swelling characteristics and used it as seed. They then polymerized it in the presence of styrene monomer and (using an electron micrograph technique) observed a cross-linked core with a grafted polystyrene shell.

Grabowski at Borg-Warner [5] patented compositions of an emulsion ABS polymer, produced by grafting styrene and acrylonitrile onto a polybutadiene latex core, and blends with polycarbonates. W. C. Calvert [6] was awarded patents for ABS blends in 1959 and for graft terpolymers in 1966.

Molau and Keskkula [7,8] studied the rubber particle formation during polymerization in high-impact polystyrene (HIPS) with and without agitation during the mass or solution polymerization process. In the case of agitation, as the ratio of polystyrene to rubber increases, a point is reached where phase inversion occurs. At this point in the reaction, the swollen rubber phase is reduced to discrete particles dispersed in the continuous polystyrene solution. Rubber-polystyrene graft copolymer molecules act as surface-active agents and stabilize the polymeric oil in oil emulsion. The final product (HIPS) has a polystyrene continuous phase and dispersed rubber particles.

Williams et al. [9,10,11] reported that in emulsion polymerization of styrene or styrene-butadiene, the polymer formed in later stages is located preferentially near the particle surface. From this observation, Williams suggests the equilibrium swelling of latex particles is not uniform, but has a core–shell conformation, the outer layer having a high monomer concentration and the interior core being polymeric. The concentration of polymer at the water interface is low and the concentration of monomer at the interface is high because of the low water–polymer solubility versus water–monomer solubility.

Kato et al. [12] claim in a U.S. patent that a weatherable, impact-resistant multilayered graft–copolymer particle for toughening rigid resins can be prepared in two or more steps starting with a crosslinked elastomeric latex particle. These acrylic core particles have a degree of swelling of 7 to 30 versus 7 to 3 for the outer layer. If the degree of swelling for the core is greater than 30, the molded articles show unfavorable appearance, while the impact strength is too low if the degree of swelling is lower than 7. The degree of swelling of the copolymers formed in the outer layers and later stages are selected from the range of 7 to 3 chiefly for the purpose of improving the appearance of the molded articles.

Dickie and Newman [13,14,15] reported the use and preparation of graded-rubber particles having hydroxyl functional-

ity at the particle surface. These special particles were designed to improve impact strength and fatigue properties of
thermosets. The graded-rubber particles are characterized by
having a rubber core, a glasslike polymeric outer shell, and hy-
droxy surface functionality. The thermosettable prepolymers illustrated for use with these hydroxyl-functional graded-rubber
particles are carboxy-functional resins, epoxy-functional resins, hydroxy-functional resins, melamine-formaldehyde resins, and novolac resins. The particulate "graded-rubber" elastomeric core is a cross-linked acrylic rubber with a soluble cap
outer shell and these particles are stable to coagulation when
melt blended.

Fahrenholtz and Kwei [16] observed that certain substi-
tuted novolac resins are compatible with a wide range of polymers. It is important to "match" polarities of polymers in order to have compatible mixtures. Even polymers containing
groups that have a high tendency to form hydrogen bonds,
such as polyphenols and polymethacrylates, will not be compatible unless modified so that the interaction between unlike
molecules becomes more favorable than the interaction between like molecules. Modifying the novolacs with alkyl
substituents and modifying polystyrene with acrylonitrile
(SAN) are examples of this polarity control that results in
compatible mixtures. This modification is also important for
GRCs where the shell polymers are grafted to the core rubber
and need to be compatible with the continuous phase to promote particle dispersion stability.

Stutman et al. [17] studied polybutyl acrylate/polystyrene
(PBA/PS) core–shell latex polymer morphology. The rubber
core and glassy shell types have been known to yield improved
impact strength. The goal of this research was to determine
which process parameters influence grafting and particle morphology using a PBA seed procedure from a recipe adopted
from Dickie et al. [13,14,15]. It is proposed that the aqueous-
phase polymerization leads to secondary PS particle formation
and that the predominant polymerization mechanism is a bal-

ance between capture of the newly nucleated particles and polymerization in a monomer-rich PBA seed particle surface layer.

Vanderhoff et al. [18] studied the emulsion polymerization of 80:20 styrene-acrylonitrile monomers in the presence of 190-nm and 300-nm monodisperse polystyrene seed latexes. Batch, batch-with-equilibrium swelling, and semicontinuous polymerizations in the presence of polystyrene (PS) seeds were used to determine the degree of grafting for each system. The Arrhenius plot of log polymerization rate of the seeded batch copolymerizations with reciprocal temperature comprised two linear regions with a sharp inflection point at 65°C. Specific volume measurements showed that the glass temperature (T_g) of the monomer-swollen SAN was also 65°C. The final latex comprised the original seed particles grown to a larger size, and in some cases a new crop of particles formed during the polymerization. The critical factor determining the formation of new particles was surface area of the seed latex; at 226 m^2/dL, new particles were not formed; at 179 m^2/dL a new crop of particles was nucleated, the number increasing with decreasing surface area. The degree of grafting of the PS seed substrate was greater for the smaller particle size seed latex and increased exponentially with increasing seed surface area. The amount of grafted SAN determined the stability of the grafted particles in acetone, a good solvent for SAN copolymer. Dynamic mechanical spectroscopy shows that the continuous phase was either the PS substrate $(T_g = 104°C)$ or the SAN phase $(T_g = 120°C)$ except where the degree of grafting was high, in which case the T_g was intermediate between the two values. Electron microscopy of thin film sections stained with RuO$_4$ confirmed which phase was continuous and showed interpenetrating networks for those systems of intermediate T_g.

Methyl methacrylate grafted rubber concentrates (GRC-MMA) have been found to be useful for modifying polystyrene, HIPS, and incompatible blends of ABS and HIPS. The

modification of HIPS with a GRC-MMA is surprising since
PMMA and PS are not compatible. Modification of styrenics
using GRC-MMA is covered by the following Dow patents:
U.S. 4,366,289 [37], 4,460,744 [59], and 4,508,871 [60].
Merkel et al. [19] studied PMMA grafting reactions on
polybutadiene-seeded latexes. The graft polymerization of
MMA on polybutadiene emulsion latexes proceeded via a
chain-transfer mechanism involving hydrogen abstraction in
the alpha position on the polybutadiene backbone. The surface area of the particle increased linearly with increasing
grafting. The grafting level decreased with increasing shell
thickness and initiator concentration.

B.   Particle Size Control

Synthetic rubber latex preparation occured in 1927, by heating a mixture of monomer, water, soap, and hydrogen peroxide
[61]. It was initially thought that polymerization occured in
the emulsified monomer droplets, hence the term emulsion
polymerization. Fikentscher [62] first suggested, in 1938, that
the polymerization actually occurs in the aqueous phase with
the monomer droplets serving as supply reservoirs. During
World War II, the U.S. government set up the Rubber Reserve
Company to: (1) stockpile and locate sources of natural rubber
and (2) develop a synthetic rubber industry. A joint research
program composed of industries, universities, and institutes
selected a recipe for rubber latex production. The 1942 recipe
was called the GR-S Recipe, Mutual Recipe, Standard Recipe,
or Harkins Mechanism (see Table 13). Harkins [20] later published a general theory of emulsion polymerization in 1947.
This type of latex provides the basic rubber particles that
when subsequently grafted with a variety of monomers, become grafted rubber concentrates (GRC).

   This GR-S latex was coagulated with dilute sulfuric acid
or aluminum sulfate to form a crumb, which was washed and
dried. Willis [21] studied SB emulsion polymerizations with-

Table 13 GR-S Recipe[a]

Butadiene	75.0
Styrene	25.0
Water	180
Soap	5.0
Potassium persulfate	0.3
Dodecanethiol	0.5

[a]Temperature, 50°C; rate (%/hr), 6; short stop at 75% conversion with hydroquinone.

Table 14 Willis's "No Emulsifier" Recipe[a]

Butadiene	75.0
Styrene	25.0
Water	200
Persulfate	1.6
Metaphosphate	1.5

[a]pH, 3.6; temperature, 50°C; time, > 100 hr.

out emulsifier using the very long polymerization time shown in Table 14. In this process large particles were made with limited coalescence at low conversion levels. Rate of polymerization was increased by using a redox system of persulfate and bisulfite and by replacing some of the styrene with acrylonitrile. The use of the water-soluble comonomer helps water-phase polymerization.

In 1954 Alfrey et al. [22] reported the preparation of large uniform polystyrene and polyvinyltoluene particles in the 100 to 1000 mμ range using latex polymerization techniques. They, along with Vitkuske and colleagues [23], then studied the effect of emulsifier level in seeded emulsion polymerization of polyvinyltoluene. They found that the preparation of uniform particles requires control of the emulsifier level so that new particles are not created during the reaction but enough emulsifier is used to prevent agglomeration. When the process of starting with seed particles at 500 Å and growing them to a final size of 2,500 Å, emulsifier is added during the reaction as the total particle surface area increases.

In producing large rubber particles for foam rubber, Daniels et al. [24] abandoned the route of direct polymerization in favor of "partial agglomeration." Howland et al. [25] listed the various methods of partial agglomeration or limited coalescence:

1. Addition of acetone (reduces dielectric constant of water phase).
2. Addition of Toluene (swells particles and force them together).
3. Neutralization of soap with acid (reduces surface charge κ).
4. Addition of electrolyte (increases κ, because its reciprocal is a measure of the diffuse ion atmosphere around the particle).
5. Freezing (the Talalay or Freeze Agglomeration Process, British Patent 758,662, involves rapidly freezing a small particle latex on a rotating drum; the ice chips are removed and quickly thawed).
6. Pressure agglomeration (forces latex through orifice at high rate).
7. Addition of defective surfactants or certain water-soluble polymers.

To obtain the large particles required to reinforce polystyrene, White et al. [26] demonstrated limited coalescence with cold-rubber styrene-butadiene (SB) latex, made at 5°C, to about 80% conversion with only about 34% soap coverage. They added polyvinyl methyl ether and K_2SO_4 as electrolyte at 50°C and observed some of the particles aggregating and coalescing to form larger spheres with a decrease in overall total surface area. Particle size analysis showed a bimodal population with a decrease in the original population of particles and an increase in population of coalesced spheres.

At about this time, Keppler et al. [27] reported on the process of limited agglomeration of latexes by latexes. One percent of low T_g hydrophilic latex of ethyl acrylate–acrylic acid (95:5) was added to 99% of low T_g hydrophobic latex of polybutyl acrylate of about 900 Å. The pH was adjusted to 8.5, and agglomeration of particles to $D_n = 6300$ Å and $D_w = 7300$ Å resulted.

Chung-li et al. [28] prepared monodisperse latexes in diameters ranging from 1 to 4 μm and studied mechanisms. They suggest that new small particles swollen with monomer are formed in the aqueous phase and then heterocoagulate with seed particles, and in turn carry monomer to seed particles. At low seed concentrations, the tiny new particles coagulate with each other, building a crop of new growing particles. At high seed concentrations, the tiny new particles are swept up by the seed particles, permitting the generation of large monodisperse emulsifier-free polystyrene latexes.

Dupre [29] claimed a method for preparing improved monoalkenyl aromatic polyblends having discrete bimodal rubber particles as the dispersed phase using a continuous mass polymerizing process. He also included methods for determining % gel, % graft, and swelling index.

Moore and Lefevre [30] used a method of isolating solids from a latex polymerized with a pH-sensitive emulsifier by passing carbon dioxide through the latex to lower the pH, thereby destabilizing the latex, and then subjecting the latex to high shear. This shear coagulation technique can be used broadly for many latex compositions.

Henton and O'Brien [31] claim in a patent that the particle size distribution in elastomeric latex preparation is controlled through an agglomerating agent. The new agglomerated and grafted particles can be recovered and used in polymer blends while retaining the favorable larger particle size distribution.

C. ABS and Toughened Materials

Once the GRCs are prepared, they are incorporated into a matrix that is lacking in toughness. For a review of the processes see Chapter 8. We now consider some more specific contributions to ABS technology.

Czerwinski [32] patented a method for blending acrylonitrile-styrene-butadiene solvent-soluble graft copolymers and epoxy resins in a solvent or mixing these primary components without solvent on a rubber mill. Finally, a curing agent is added prior to utilization as films, fibers, adhesives, coatings, and so on.

Fromuth and Shell [33] claim high-impact and solvent-resistant compositions comprising (1) about 25 to 95 wt % of an aromatic polyester, (2) about 1 to 8% of an aromatic polycarbonate, and (3) the balance of butadiene-based core–shell particles (graft copolymer). Preferably more than one vinyl monomer is grafted to the butadiene rubber; for example, a preferred graft copolymer is a three-stage polymer having a butadiene-based rubbery core, a second stage polymerized from styrene, and a final stage, or shell, polymerized from methyl methacrylate and 1,3-butylene glycol dimethacrylate.

Bredeweg [34] produced a mass-type ABS by grafting SAN copolymer onto the polybutadiene-rubber backbone. The rubber phase may also be a SB rubber.

Morris et al. [35] have patented a way of producing ABS-type polyblend compositions with improved physical properties. Compositions contain from about 5 to 60% by weight of a copolymer of styrene and acrylonitrile; from about 10 to 70% by weight of a graft copolymer of styrene, acrylonitrile, and a butadiene polymer having an average rubber particle size of about 1 to 10 μm; and from about 10 to 70% of a high butadiene content backbone copolymer of styrene, acrylonitrile, and a butadiene polymer having an average rubber particle size of less than about 0.5 μm. Suitable large rubber particles can be prepared by mass suspension or mass polymerization techniques. The small rubber particles are made by emulsion polymerization. The polyblend is composed of a mixture of two ABS resins, one of low butadiene content (< 10 wt % B) of large particles and the other of small rubber particles with high butadiene content (> 50% B) and a thermoplastic SAN copolymer. See Table 15 for possible composition ranges for the rubber particles.

Dupre [36] was able to increase the toughness of HIPS polyblends by 50%. This was done by using a bimodal rubber particle distribution with 70 to 95% of the particles from 0.5 to 1.0 μm in size and 5 to 30% of the rubber particles from 2 to 5 μm on the average.

Keskkula et al. [37] found that acrylate-grafted elastomers are useful as polymer modifiers in melt blending with styrenics (homopolymer, copolymers (SAN, SMA, SAA), block copolymers (S-I-S type), and graft copolymers (HIPS), and optionally ABS or ABS-type resins].

Falk et al. [38] claimed the use of thermoplastic core–shell compositions having a rigid core, a rubbery acrylic shell, and a transition layer between the core and shell layers. Although the rigid and rubbery components include interactive functional monomers, the resulting elastomers remain thermally processable in extrusion and injection molding equipment.

Table 15 Rubber Particles of U.S. Patent 4,250,271

Polyblend (wt %)	Description	Rubber particle size (μm)
5–45	Thermoplastic SAN	
25–60	Graft copolymer with < 10% B made by mass suspension or mass polymerization	1–10
20–50	Backbone copolymer with > 50% B made by emulsion polymerization	< 0.5

Source: Ref. 35.

Meunier [39] claimed impact improvement in a thermoplastic polymer by the addition of a graft copolymer comprising a backbone copolymer consisting of 0.5 to 35 wt % of butadiene or isoprene and an alkyl acrylate of C_2 to C_{12} alkyl and optionally a polyfunctional cross-linking agent, onto which is grafted a copolymer consisting of a C_1 to C_4 alkyl methacrylate and a C_1 to C_8 alkyl acrylate. The resin compositions are particularly useful to improve impact strength at low temperatures.

Koster and Traugott [40] have developed polymer blends of transparent impact polystyrene (TIPS). At 10% polybutadiene, the styrene–methyl methacrylate ratio is adjusted to give a refractive index that approaches that of polybutadiene. If additional toughness is desired, acrylonitrile can be added, producing transparent ABS (TABS) [41].

Schlund and Lambla [42] demonstrated that it is possible to coat glass fibers with a thick interlayer of a coupling agent and a reactive latex, using a single-step, on-line industrial process. They also show that a low T_g, reactive latex, by forming a core-reactive shell by "stage-polymerization" gave desirable composite properties.

Fowler et al. [43] observed an order-of-addition protocol for location of nonreactive surfaces of MBS emulsion particles in immiscible PS-SAN blends. It was found that the sequence of mixing had a strong influence on the location of the MBS particles. If the PS-SAN interface is established before the addition of the MBS particles, the MBS particles are located exclusively in the SAN phase. If the MBS particles are present at the time the PS-SAN interface is formed, then the particles line up at the interface.

D. Other Materials

A weatherable impact modifier for PVC, Acryloid KM-323B (modified butyl acrylate rubber with 20% methyl methacrylate) was introduced in the mid-1970s [44].

Chung et al. [45] claimed a thermoplastic blend comprising a C_1 to C_4 (alkylene terephthalate), a polycarbonate resin, and a butadiene- or an acrylate-based graft copolymer to be amenable without pre-extrusion drying, to thermoplastic processing upon incorporation of an additive amount of a polymeric modifier.

Sederel [46] found that polymer mixtures containing an aromatic polycarbonate resin, a block copolymer (hydrogenated S-B-S), and a graft copolymer (MBS) have good impact strength at low temperatures, and do not show any delamination phenomena when injection molded.

Rifi et al. [47] showed that end-use performance of impact polypropylene depends on crystallinity and molecular weight of homopolymer and modifiers, size and distribution of modifier domains, and concentrations of modifiers. Tests show optimum property balances are achieved via in-reactor operation, rather than in postblending procedures. An impact modifier must allow for dissipation of the stress created by the impact. Blends with large-particle rubber domains of ≈ 6 μm yield poor impact, but domains < 1 μm yield good impact.

Nakamura et al. [48,49] found that submicrometer acrylic core–shell latex particles were effective in reducing the internal stress in a cured epoxy resin system. The powdered core–shell particles were dispersed in the resin before the curing process. The key parameters for the rubber particles for reducing internal stress in the cured system were core–shell particle diameter, shell thickness and hardness, crosslink density of the core rubber, and chemical reactivity of the particle surface with the epoxy matrix.

The toughening mechanisms in a core–shell rubber-toughened brittle epoxy system have been studied by Sue [63] using transmitted light optical microscopy and transmission electron microscopy techniques. The toughening of this modified, brittle epoxy system was found to be due to cavitation of the core–shell rubber particles, followed by formation of limited shear banding when the crack propagates, rather than the expected crack–particle bridging mechanism.

Udipi et al. [50] observed core–shell particles to be useful in producing rubber-toughened Nylon 6 by reaction injection molding. The core–shell rubber latex is mixed with aqueous caprolactam monomer and the water removed to form a colloidal NAD of core–shell rubber in the monomer. The colloidal rubber remains stable even during caprolactam polymerization. Incorporation of a postreactive functional hydroxyl group on the rubber particle surface brings about further improvement in impact strength. Bonding the discrete particles to the matrix appears to give the optimum rubber toughening for thermoplastic Nylon 6 polymer.

Meredith and Ferguson [51] formed clear modified PVC by blending a graft copolymer including from 60 to 70 wt % of a butadiene-based rubber core and a grafted SAN copolymer shell with S:AN ratios of 3/1 to 5/1. These compositions have excellent clarity and good impact properties.

Falk and Kliever [52] claim that blends of thermoplastic nitrile rubber graft copolymers, PVC, and plasticizer are useful as thermoplastic elastomers. The compositions do not require vulcanization, do exhibit a rubbery feel and appearance, have good resistance to compression set, and may be melt processed in conventional molding and extrusion equipment.

Henton et al. [53] describe specific particulate rubber dispersions that are stable and useful as curable coatings and toughened resins for structural composites. Similar compositions are also useful for fiber-reinforced laminates; advanced composites useful in aerospace structures, fiberglass-reinforced plastics tooling, casting and molding resins; bonding agents; adhesives; and encapsulants of electrical components [48,49] that are exposed to wide temperature fluctuations. GRC-type rubbers with crosslinked cores and graftable sites are insoluble in the polyepoxide resin, but the rubber is sterically stabilized with a postreactive grafted shell to allow particle transfer from a dispersion in water to a stable dispersion in the reactive resin, while remaining curable with the resin forming a strong interfacial-bonding rubber-toughened epoxy resin thermoset. The shell components contain a functionality that reacts with epoxy resin phase to impart toughening to the continuous phase, as also observed by Udipi [50].

Sue [56] found toughness measurements using various modified epoxy systems, including core–shell rubber-toughened brittle epoxy, produced correlations between the single-edge-notch three-point bend, the single-edge-notch four-point bend, and the double-notch four-point bend methods. The toughness of both the neat and rubber-toughened epoxy systems were found to be independent of the testing techniques used.

Yang and Pickelman [54] found that polytriazines based on cyanate ester resins exhibit improved toughness without sacrificing heat resistance when prepared by curing in the presence of GRC particles grafted with a resin-soluble shell polymer. The resin-soluble shell sterically stabilizes the rubber particles and maintains the particles in an essentially discrete, nonagglomerated form. These core–shell rubbers are effective in toughening the brittle polycyanate ester–cured resin. Unlike a soluble rubber system (VTBN), its morphology is predetermined and does not lead to T_g depression. Rubber-modified polycyanate with VTBN and Proteus 5025, rubbers from BF Goodrich, and GRC core–shell rubber were evaluated and compared to the unmodified control polycyanate. Fracture toughness results from compact tension [55] are given in Table 16. The core–shell modified system gives the best toughness improvement without the usual loss in T_g.

Sue [56] evaluated particulate modified GRC preformed rubbers in toughening a brittle epoxy resin cure system DGEBA/DDS (diaminodiphenyl sulfone). The GRC rubber provided the most effective toughening, followed by preformed particles and then Proteus rubber. A summary of the fracture toughness of the various epoxy systems is shown in Table 17.

Sue et al. [58] evaluated the chemistry of designed GRC core–shell rubbers with the brittle epoxy resin cure system DGEBA/DDS. Particle shell thickness (core–shell ratio), surface polarity (acrylonitrile content), and surface latent coreactivity (glycidyl methacrylate) were varied using seven compositions. The surface chemistry varied, but the core rubber

Table 16 Fracture Toughness of Cyanate XU71787 with 10% Rubber Modification

Rubber	K_{IC}(initiation) MPa $\times$ m$^{0.5}$	T_g(°C)
No rubber	—	—
XU71787[a]	0.53	254
VTBN[b]	0.72	235
Proteus 5025[c]	0.69	—
CSR F[d]	1.36	254

[a] A curable resin based on Polyfunctional aromatic polycyanates and dicyclopentadiene.

[b] A reactive liquid rubber with terminal vinyl groups from BF Goodrich.

[c] Proteus™ 5025, a rubber from BF Goodrich.

[d] A core–shell rubber described in U.S. Patent 4,894,414. XU71787 is an experimental cyanate ester resin with 10% colloidal core–shell rubber offered by The Dow Chemical Company.

Source: Ref. 55.

Table 17 Summary of Fracture Toughness of Various Epoxy Systems

Material	T_g (°C)	K_{1c}[f] (MPa $\times$ m$^{1/2}$)	G_{1c}[g] (J/m^2)
DER®332[a]/DDS[b]	220	0.84	182
DER®332/GRC[c]/DDS	219	1.20	496
DER®332/DAR[d]/DDS	197	1.08	422
DER®332/Proteus™[e]/DDS	214	0.97	303

[a] Epoxy resin product of The Dow Chemical Company.

[b] Curing agent diaminodiphenyl sulfone.

[c] Core–shell rubber described in U.S. 4,778,851 (1988) [53].

[d] Nonaqueous dispersion of a dispersed polyacrylate rubber described in U.S. 4,708,996 (1987) [57]; 4,789,712 (1988) [57].

[e] Rubber from BF Goodrich.

[f] Fracture toughness defined as stress intensity factor.

[g] Fracture energy as measured by the double-notched four-point-bend (DN-4PB) technique.

Source: Ref. 56.

was held constant. Four monomers chosen for surface modification of GRC were styrene (S), methyl methacrylate (MMA), acrylonitrile (AN), and glycidyl methacrylate (GMA). The appropriate level of GMA helps disperse the particles more randomly. A minimum AN level is required to prevent particle agglomeration. Increasing shell thickness adds a crack deflection mechanism to the modified GRC particle that already exhibits cavitation and matrix shear yield mechanisms. Control of particle morphology and chemistry can aid the modified GRC particle cavitation, the matrix shear yield mechanism, and the crack deflection mechanism to produce a synergistic toughening effect. A summary of toughness and T_g of these designed variations is given in Table 18.

Table 18 Summary of K_{IC} and T_g of Designed GRC-Modified Rubber in a Brittle DER[a]332/DDS[b] System

Core/ shell (wt ratio)	Shell (S/MMA/AN/GMA)[d]	K_{IC} (MPA × m$^{1/2}$) GRC-modified system[c]	T_g (°C)
Unmodified		0.83	200
A. 84/16	22.5/22.5/25/30	1.20	222
B. 84/16	30/30/25/15	1.20	219
C. 84/16	37.5/37.5/25/0	1.30	223
D. 84/16	43.8/43.7/12.5/0	1.07	229
E. 84/16	50/50/0/0	1.05	224
F. 75/25	30/30/25/15	1.33	223
G. 65/35	30/30/25/15	1.37	220

[a] Epoxy resin product of The Dow Chemical Company.
[b] Curing agent diaminodiphenyl sulfone.
[c] Core–shell rubber based on U.S. 4,778,851 (1988) [53].
[d] Shell-copolymers of styrenemethyl methacrylate/acrylonitrile/glycidyl methacrylate.
Source: Ref. 58.

III. SUMMARY

The GRC core–shell rubber modifier has versatility and characteristics, both physical and chemical, for toughening almost any polymer system, either thermoplastic or thermoset. Composition changes, process improvements, methods of polymer recovery, and blend operations of GRCs are all important in developing toughened polymers. Understanding failure mechanisms and determining structure–property relations for toughened systems will make it possible to predict which GRCs will be most effective.

REFERENCES

1. T. Patton, *Alkyd Resin Technology*, Interscience, New York, 1962.
2. R. Seymour R, Origin and Early Development of Rubber-Toughened Plastics, *Rubber-Toughened Plastics* (C. K. Riew, ed.), Advances in Chemistry Series 222, American Chemical Society, Washington, DC, 1989, pp. 3–13.
3. C. Childers and C. Fisk (to U.S. Rubber Company), U.S. Patent 2,820,773, 1958.
4. E. Bradford and J. Vanderhoff, *J. Polym. Sci. C.* 3:41 (1963).
5. T. Grabowski (to Borg-Warner), U.S. Patent 3,130,177, 1964.
6. W. Calvert (to Borg-Warner), U.S. Patents 2,908,991, 1959, and 3,238,275, 1966.
7. G. Molau and H. Keskkula, *Appl. Polym. Sympos.* 7:35 (1968).
8. G. Molau and H. Keskkula, *J. Polym. Sci. A-1,4*:1595 (1968).
9. M. Grancio and D. Williams, *J. Polym. Sci. A-1,8*:2617 (1970).
10. D. Williams, *J. Polym. Sci. Chem. 11*:301 (1973).
11. P. Keusch and D. Williams, *J. Polym. Sci. Chem. 11*:143 (1973).
12. T. Kato, M. Izuml, K. Chikanishi, R. Handa, J. Kobayashi, Weather and Impact-Resistant Resin Composition Comprising a Graft Copolymer Containing Multi-Layer Polymer Particles and a Rigid Resin (to Mitsubishi Rayon Co. Japan), U.S. Patent 3,830,878, 1974.

13. R. Dickie, M. Cheung, and S. Newman, *J. Appl. Polym. Sci.* *17*:65 (1973).

14. R. Dickie and S. Newman, Acrylate Polymer Particles: A Core, an Outer Shell, and an Intermediate Layer (to Ford Motor Company), U.S. Patent 3,787,522, 1974.

15. R. Dickie and S. Newman, Graded Rubber Particles Having Hydroxyl Functionality and a Polymeric Crosslinking Agent (to Ford Motor Company), U.S. Patent 3,856,883, 1974.

16. S. Fahrenholtz and T. Kwei, Compatibility of Polymer Mixtures Containing Novolac Resins, *Macromolecules 14*:1076 (1981).

17. D. Stutman, A. Klein, M. El-Aasser, and J. Vanderhoff, Mechanism of Core/Shell Emulsion Polymerization, *Ind. Eng. Chem. Prod. Res. Dev. 24*:404 (1985).

18. J. Vanderhoff, V. Dimonie, M. El-Aasser, and A. Klein, Emulsion Copolymerization of Azeotropic Styrene–Acrylonitrile Monomer Mixture in Polystyrene Seed Latexes, *Makromol. Chem. Suppl. [MCSUEU] 10–11*:391 (1985).

19. M. Merkel, V. Dimonie, M. El-Aasser, and J. Vandehoff, Poly(methylmethacrylate) Grafting Reactions Inside Polybutadiene Seeded Latexes, *Polym. Mater. Sci. Eng. [PMSEDG] 1* (*54*):598 (1986).

20. W. Harkins, *J. Am. Chem. Soc. 69*:1428 (1947).

21. J. Willis, *Ind. Eng. Chem. 41*:2272 (1949).

22. T. Alfrey, Jr., E. Bradford, and J. Vanderhoff, *J. Op. Soc. Am. 44*:603 (1954).

23. J. Vanderhoff, J. Vitkuske, E. Bradford, and T. Alfrey, Jr., *J. Polym. Sci. 20*:225 (1956).

24. T. Daniels, W. Watson, and F. White, *Rubber Plast. Age 40*:1057 (1959).

25. L. Howland, E. Aleksa, R. Brown, and E. Borg, *Rubber Plast. Age 42*:870 (1961).

26. W. White, J. Reynolds, and R. Gilbert, *J. Appl. Polym. Sci. 8*:2049 (1964).

27. H. von Keppler, H. Wesslau, and J. Stabenow, *Angew. Macromol. Chem. 2*:1 (1968).

28. Y. Chung-li, J. Goodwin, and R. Ottewill, *Progr. Coll. & Polym. Sci. 60*:163 (1976).

29. C. Dupre, Method for Preparing a Monoalkenyl Aromatic Polyblend Having a Dispersed Rubber Phase as Particles with

a Bimodal Particle Size Distribution (to Monsanto Co.), U.S. Patent 4,146,589, 1979.

30. E. Moore and N. Lefevre, Method for Shear Coagulation of Latex Resins (to The Dow Chemical Company), U.S. Patent 4,623,678, 1986.

31. D. Henton and T. O'Brien, Particle Agglomeration in Rubber Latices (to The Dow Chemical Company), U.S. Patent 4,419,496, 1983.

32. R. Czerwinski, Blends of Epoxy Resins and ABS Graft Copolymers (to Borg-Warner), U.S. Patent 3,496,250, 1970.

33. H. Fromuth and K. Shell (to Rohm & Haas Co.), U.S. Patent 4,180,494, 1979.

34. C. Bredeweg, (to The Dow Chemical Company), U.S. Patent 4,239,863, 1980.

35. J. Morris, J. Brandli, E. Lanza and W. Lingier, ABS-Type Polyblend Compositions (to Cosden Technology), U.S. Patent 4,250,271, 1981.

36. C. Dupre, Process for Preparing Polymeric Polyblends Having a Rubber Phase As Particles with a Bimodal Particle Size Distribution (to Monsanto Co.), U.S. Patent 4,334,039, 1982.

37. H. Keskkula, D. Maass, and K. McCreedy, Acrylate-Grafted Elastomers as Polymer Modifiers (to The Dow Chemical Company), U.S. Patent 4,366,289, 1982.

38. J. Falk, S. Mylonakis, and D. VanBeek, Thermoplastic Acrylic Elastomers (to Borg-Warner), U.S. Patent 4,473,679, 1984.

39. G. Meunier, Graft Copolymers of Alkyl Methacrylates- Alkyl Acrylates onto Diene-Alkyl Acrylate Copolymers and Their Use as Impact Modifiers, (to M&T Chemicals) U.S. Patent 4,542,185, 1985.

40. R. Koster and T. Traugott (to The Dow Chemical Company), U.S. Patent 4,530,973, 1985.

41. E. R. Moore, ed., Styrene Polymers, *Encyclopedia of Polymer Science & Engineering*, Vol. 16, 1989, p. 75.

42. B. Schlund and M. Lambla, Modification of the Damping Behavior of Epoxy/Glass Fiber Composites via Fiber Coating with Functional Latices, *Polym. Compos.* 6(October):272 1985.

43. M. Fowler, H. Keskkula, and D. Paul, Distribution of MBS Emulsion Particles in Immiscible Polystyrene/SAN Blends, *J. Appli. Polym. Sci.* 37:225 (1989).

44. Modifier Performance Properties Acryloid® KM-323B, weatherable high efficiency impact modifier for PVC, *Rohm & Haas Product Bulletin MR-94*, 1980.
45. J. Chung, R. Carter, and D. Neuray, Thermoplastic Resinous Blend and a Method for Its Preparation (to Mobay), U.S. Patent 4,554,315, 1985.
46. W. Sederel, Polymer Mixture Which Comprises an Aromatic Polycarbonate Resin and an Impact Strength--Improving Agent (to General Electric Company), U.S. Patent 4,617,345, 1986.
47. M. Rifi, H. Ficker, and D. Walker, How in situ modifier production improves impact properties of polypropylene, *Mod. Plast.* 64:(February):62, 1987.
48. Y. Nakamura, H. Tabata, H. Suzuki, K. Iko, M. Okubo, and T. Matsumoto, Internal Stress of Epoxy Resin Modified With Acrylic Core--Shell Particles Prepared by Seeded Emulsion Polymerization. *J. Appl. Polym. Sci. 32*:4865 (1986).
49. Y. Nakamura, H. Tabata, H. Suzuki, K. Iko, M. Okubo, and T. Matsumoto, Internal Stress of Epoxy Resin Modified with Acrylic Core--Shell Particles Containing Functional Groups Prepared by Seeded Emulsion Polymerization, *J. Appl. Polym. Sci. 33*:885 (1987).
50. K. Udipi, Particulate Rubber--Modified Nylon 6 RIM, *J. Appl. Polym. Sci. 36(1988)*.
51. F. Meredith and L. Ferguson, Clear Impact Modifier for PVC (to Borg-Warner), U.S. Patent 4,764,563, 1988.
52. J. Falk, I. Kliever, Grafted Nitrile Rubber--Plasticized PVC Blends as Thermoplastic Elastomers (to Borg-Warner), U.S. Patent 4,764,552, 1988.
53. D. Henton, D. Pickelman, C. Arends, and V. Meyer, Rubber-Modified Epoxy Compounds (to The Dow Chemical Company), U.S. patent 4,778,851, 1988.
54. P. Yang and D. Pickelman, Rubber-Modified Cyanate Ester Resins and Polytriazines Derived Therefrom (to The Dow Chemical Company), U.S. Patent 4,894,414, 1990.
55. P. Yang, E. Woo, M. Bishop, D. Pickelman, and H. Sue. Rubber Toughening of Thermosets—A System Approach, *Proceedings, ACS Division of Polymeric Material, Science and Engineering 63*, 315 (1990).

56. H.-J. Sue, Study of Rubber-Modified Brittle Epoxy Systems. Part I: Fracture Toughness Measurements Using the Double-Notch Four-Point-Bend Method, *Polym. Eng. Sci. 31*:275 (1991).

57. D. Hoffman and C. Arends, Stable Dispersions In Polyepoxides (to The Dow Chemical Company), U.S. Patents 4,708,996, 1987, and 4,789,712, 1988.

58. H.-J. Sue, E. Garcia-Meitin, D. Pickelman, and P. Yang, Toughening of High Performance Epoxies Using Designed Core–Shell Rubber Particles, *ACS Polym. Preprint 32*:358 (1991).

59. H. Keskkula, D. Maass, and K. McCreedy, Blends of Grafted Acrylate Polymers and Mass-Made ABS-Type Resins (to The Dow Chemical Company, U.S. Patent 4,160,744, 1984.

60. H. Keskkula, D. Maass, and K. McCreedy, Transparent Blends of Polymethylmethacrylate and Certain Styrene Copolymers (to The Dow Chemical Company), U.S. Patent 4,508,871, 1985.

61. M. Luther and C. Heuck (to I. G. Farben Co.) German Patent DRP 558,890 (1927).

62. H. Fikentscher, *Angew. Chem. 51*:433 (1938).

63. H.-J. Sue, Study of Rubber Modified Brittle Epoxy Systems. Part I: Toughening Mechanisms Under Mode—I Fracture, *Polym. Eng. Sci. 31*(4), (1991).

4
Crazing in Intrinsically Tough-High-Performance Thermoset Resins

PHILIP C. YANG   The Dow Chemical Company, Plaquemine, Louisiana

HUNG-JUE SUE*   The Dow Chemical Company, Freeport, Texas

MATTHEW T. BISHOP   The Dow Chemical Company, Midland, Michigan

I. INTRODUCTION

Crazing, although not as effective as shear banding, is a major source of toughening in polymers [1,2]. For thermosetting polymers, however, crazing is thought to be unsuitable because of the inability of the crosslinked molecules to undergo significant

*Current affiliation: Texas A&M University, College Station, Texas

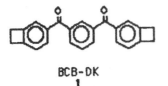

BCB-DK
1

Figure 1   Benzocyclobutene-based bis-benzocyclobutene-diketone.

molecular stretching and disentanglement, which are among the essential mechanisms for crazing to take place [1–3]. Therefore, without conclusive evidence to show otherwise, it is agreed that the crazing phenomenon in thermosets, if it occurs at all, is an abnormality [4–6]. It is well known that the typical high-performance thermoset resins are highly crosslinked and hence have high glass transition temperatures and low fracture toughness. This can be seen, for example, in the epoxy-based system DER* 383/diaminodiphenyl sulfone (DDS), the cyanate-based system dicyclopentadiene (DCPD) novolac cyanate, and bisphenol A (BIS-A) dicyanate, the benzocyclobutene-based bis-benzocyclobutene-diketone (BCB-DK) (Fig. 1), and acetylene-chromene-terminated (ACT) resin (Fig. 2). The last is a mixture of bispropargyl ether of bisphenol A (Fig. 2a), propargyl ether and chromene of bisphenol A (Fig. 2b), and dichromene of bisphenol A (Fig. 2c) and typically contains about 60% chromene and 40% propargyl ether functional groups. The chromene content can increase above 60% by further B-staging of typical ACT, becoming high-chromene (HC) ACT resin.

The fracture toughness as determined by compact tension method and T_g as determined by dynamic mechanical spectroscopy of the aforementioned polymers are listed in Table 1. It can be seen that low fracture toughness is a common feature for the high-performance thermoset resins.

Recent advancements in the synthesis of new thermosets [7–9], in the development of novel tougheners for epoxy-based resin [10,11] and cyanate-based resin [12,13], and in the use of better and more direct tools for studying fracture mechanisms

---

* Trademark of The Dow Chemical Company.

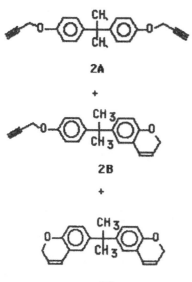

2A

+

2B

+

2C

Figure 2   Acetylene-chromene-terminated (ACT) resin.

Table 1   Properties of High-Performance Thermoset Resins

Polymer	K_{IC} (MPa m$^{1/2}$)	T_g (°C)
DER 383/DDS	0.68	242
DCPD cyanate	0.53	250
Bis-A cyanate	0.57	260
BCB-DK	0.59	340
ACT	0.49	350
HC-ACT	0.33	> 357

in polymers [14–17] have led to numerous findings of previously unreported toughening mechanisms in thermosets as well as a better understanding of why and how these toughening mechanisms operate.

The present work, benefiting from the preceding technological achievements, focuses on the toughening mechanisms of the newly developed thermoset resins, which are intrinsically tough despite high glass transition temperature

(> 200°C). The first one is based on the resins that have 1.2-
dihydrobenzocyclobutene and maleimide functional groups
(AB-BCB-MI resins) [9,18–21] and the second one is based on
the mixture of BCB-DK resin and HC-ACT resin. The work
presented here was aimed at discovering which toughening
mechanisms are operative for BCB-DK/HC-ACT and the AB-
BCB-MI polymers that presumably result in their impres-
sively high fracture toughness.

II. MATERIALS AND PROPERTIES

The synthesis and the physical and mechanical properties of a
number of AB-BCB-MI resins have been discussed by others
[9,18–21]. In many cases the fracture toughness is extremely
high, although these polymers are crosslinked and have high
glass transition temperatures ($T_g \geq 200$°C). The chemical
structures of the two AB-BCB-MI resins that were used in the
present study are shown in Fig. 3.

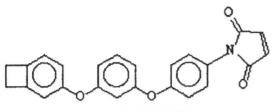

AB-BCB-MI-1

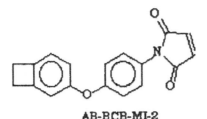

AB-BCB-MI-2

Figure 3   Chemical structure of two AB-BCBs.

Table 2 Properties of AB-BCB-MI Polymers

Polymer	T_g (°C)[a]	G' (MPa)[b]	K_{IC} (MPa m$^{1/2}$)	Swelling (%)[d]
AB-BCB-MI-1	201	2.56	3.95[c]	185
AB-BCB-MI-2A	264	1.51	2.62	171
AB-BCB-MI-2B	271	4.65	1.12	88

[a] Temperature at which the shear loss modulus (G'', at $w = 1$ rad/sec) is a maximum.
[b] Rubbery shear storage modulus ($w = 1$ rad/sec) at $T = T_g + 60$°C.
[c] Plane strain criterion not satisfied; hence, not a *critical* stress-intensity value.
[d] In methylene chloride.

Table 3 Properties of BCB-DK and ACT Blend

Polymer	T_g (°C)	G' (MPa)	K_{IC} (MPa m$^{1/2}$)	Swelling (%)
BCB-DK/HC-ACT	233	3.0	1.86	144.5
BCB-DK/ACT	292	—	0.72	30.2

The structures are similar, with 3-(oxy-4-benzocyclo-butenyl)phenyl-4'-N-maleimidophenyl ether (AB-BCB-MI-1) possessing an additional oxy-1,3-phenylene spacer as compared to 4-benzocyclobutenyl-4'-N-maleimidophenyl ether (AB-BCB-MI-2). Fracture toughness, glass transition temperature, and rubbery shear modulus for the AB-BCB-MI polymers investigated in this study are shown in Table 2.

For the 1:1 mole ratio of BCB-DK and HC-ACT blend investigated in this study, fracture toughness, glass transition temperature, rubbery shear modulus, and swelling in terms of percent weight gain of methylene chloride are shown in Table 3. For reference purposes, the properties of BCB-DK and ACT blend are given in Table 3.

III. RESULTS AND DISCUSSION

For all the high-performance thermosets shown in Tables 2 and 3, except BCB-DK/ACT, the fracture toughness is significantly higher than that of the conventional thermosets listed in Table

1. Furthermore, the presence of rubbery shear modulus and the swelling, instead of dissolving, in methylene chloride indicate these materials are thermosets, not thermoplastics.

For AB-BCB-MI polymers, given the large differences in fracture toughness among these samples, one expects corresponding differences in the size of the damage zone and perhaps also in the operative fracture mechanism. In fact, microscopic examination indicates that the fracture mechanism is essentially the same for all three AB-BCB-MI polymers, but that the plane-strain damage zone size varies tremendously. For AB-BCB-MI-1, the crazed damage zone is approximately $1800 \times 1000 \ \mu m = 1.8 \times 10^6 \ \mu m^2$ (the length ahead of the crack tip $\times$ the maximum width = cross-sectional area); for

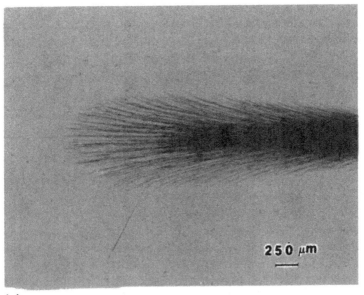

250 μm

(a)

Figure 4 Bright-field TOM micrographs of the AB-BCB-MI polymers taken in the plane strain core region of the DN-4PB crack tip damage zone: (a) AB-BCB-MI-1, (b) AB-BCB-MI-2A, and (c) AB-BCB-MI-2B. The crack propagates from right to left.

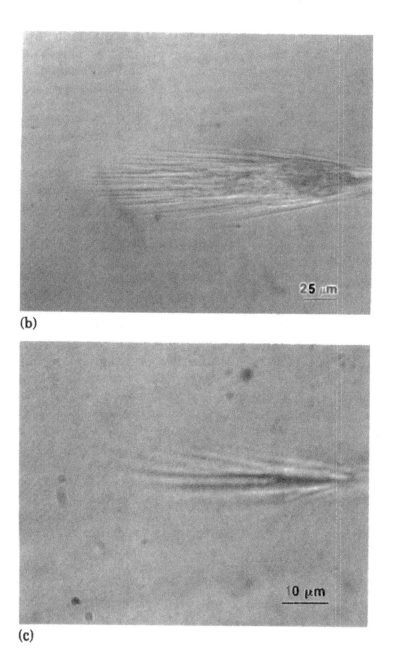

(b)

(c)

AB-BCB-MI-2A, 200×50 μm $= 1.0 \times 10^4$ μm^2; and for AB-BCB-MI-2B, 32×16 μm $= 510$ μm^2 (see Fig. 4). The cross-sectional area of craze damage for AB-BCB-MI-1 is about 20 times that for AB-BCB-MI-2A, which in turn is about 2×10^4 times that for AB-BCB-MI-2B.

Transmission optical micrographs (TOM) of the DN-4PB damage zone (Fig. 4) show that multiple cracking or crazing or both take place in all three cases. This type of damage pattern is quite different from that of ordinary thermosets, where only limited crack tip local yielding and/or occasional crack branching and bifurcation are observed [22,23]. In fact, the damage patterns shown in Figs. 4a and 4b resemble the crazelike damage we have previously observed in a core–shell rubber-modified epoxy system [6].

Visual inspection of the fracture surface created during the fracture test, that is, slow crack propagation, of AB-BCB-MI-1, which has the highest toughness, reveals a distinct layer of silvery coating. It should be noted that the color of resin by itself is light amber. The formation of a silvery layer on the fracture surface indicates the presence of a strong light-reflecting structure located on the fracture surface or slightly below it. Whereas the fracture surfaces created after fracture test by pulling apart the ligament portion of test specimen (i.e., fast crack propagation), the color of the fracture surface is the same as the resin itself. However, upon further studying of the fast fracture region of the fracture surface under reflected optical microscopy (ROM), the fracture surface is seen to be full of fracture lines with shapes similar to a fish scale (Fig. 5). It is believed that these fish scales are associated with the precursors of the secondary cracks formed ahead of the main crack or craze.

The distinct silvery layer described above does not appear in fracture surfaces of the AB-BCB-MI-2A and BCB-DK/HC-ACT polymer. Both have lower fracture toughness as compared with AB-BCB-MI-1. However, upon careful examination, the fracture surface, created by slow crack propagation, is seen to have a hazy appearance. The hazy appearance indi-

Figure 5 Bright-field ROM micrograph of AB-BCB-MI-1 polymer taken at the fast crack propagation region. The crack propagates from right to left.

cates the presence of some structure that causes the interference of reflected light. The optical micrograph taken under the cross-polarized light is shown in Fig. 6. The multicolors of the micrograph indicates the interference of reflected light caused by the structure present at the fracture surface. For reference purpose, the cross-polarized ROM of brittle material (i.e., BCB-DK/ACT) is shown in Fig. 7. The mono/chromatic appearance indicates the featureless fracture surface that is typically seen in brittle material. Interestingly, secondary cracks (Fig. 4) are also observed, in both AB-BCB-MI-2A and BCB-DK/HC-ACT resins, at the surface created by fast crack propagation.

It is intriguing to see the common features shared by all the intrinsically tough thermosets despite the difference in chemical structure among the AB-BCB-MI resins and BCB-

Figure 6 Cross-polarized light micrograph of AB-BCB-MI-2A taken at the slow crack propagation region. The crack propagates from right to left.

DK/HC-ACT resin. In order to better define the toughening mechanism in these intrinsically tough thermosets—in particular, AB-BCB-MI polymers—TEM investigations of the DN-4PB damage zones were also conducted.

The TEM micrographs from the DN-4PB crack tip damage zone of AB-BCB-MI-1 clearly demonstrate the existence of widespread and fully grown crazes (Figs. 8 and 9). The matrix immediately in front of the crack tip is highly stretched (see arrow in Fig. 8). Blunting of the crack can readily be seen. A few micrometers away from the crack tip, fully grown crazes are observed (Fig. 9). These crazes exhibit sharp boundaries with the matrix and are bridged by thin craze fibrils. Some of

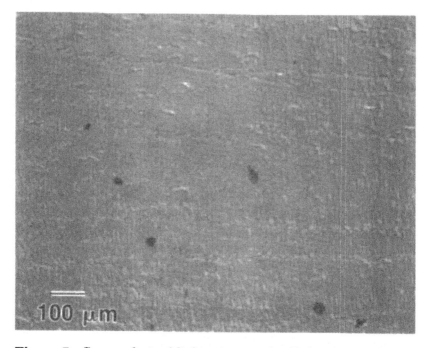

Figure 7 Cross-polarized light micrograph of BCB-DK/ACT taken at the slow crack propagation region. The crack propagates from right to left.

the crazes also contain partially grown cracks. Still farther away from the crack tip, crazes initiate and grow around some sort of inclusion or impurity (Fig. 10). This resembles the crazing phenomenon in rubber-toughened-polystyrene [1].

For a given monomer, an increase in crosslinking reduces the extent of this crazing phenomenon, which in turn leads to a decrease in fracture toughness. AB-BCB-MI-2B has a higher rubbery modulus than AB-BCB-MI-2A, which indicates that it is more highly crosslinked. Correspondingly, AB-BCB-MI-2B has scarce, yet big crazes in the crack tip damage zone (Figs. 3 and 12), as compared to AB-BCB-MI-2A, which shows relatively widespread crazes around the crack tip region (Figs.

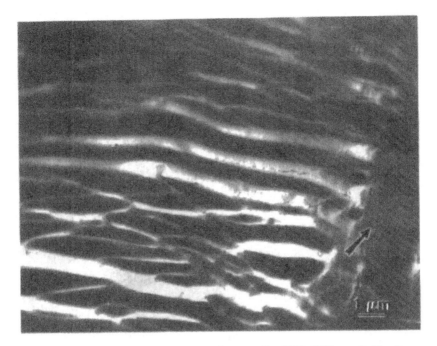

Figure 8 A TEM micrograph taken at the DN-4PB crack tip damage zone of AB-BCB-MI-1. Aside from the occurrence of extensive crazing in the damage zone, significant crack tip blunting is also observed (see arrow). The crack propagates from right to left.

3b and 11). Although these AB-BCB-MI-2 samples show the same crazing phenomenon (Figs. 8 and 9) as seen for AB-BCB-MI-1, it is not nearly as extensive.

These results suggest that, starting from a given AB-BCB-MI monomer, the lower the crosslink density of the resulting polymer, the more prone it is toward formation of stable large-scale craze zone. This conjecture has some support from Henkee and Kramer's investigation of crazing in crosslinked PS [24], where they observed that an increase in crosslinking tends to suppress crazing. However, other factors also appear to be important in leading to different amounts of crazing for different

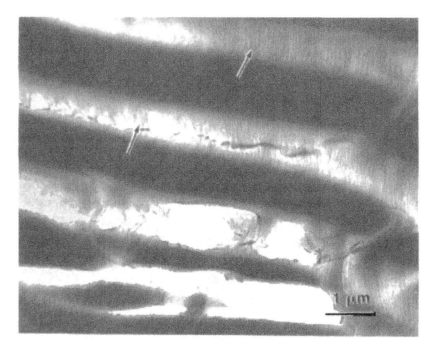

Figure 9 A TEM micrograph taken a few micrometers ahead of the crack tip of AB-BCB-MI-1. Craze fibrillation (see arrows) is clearly observed. The crack propagates from right to left.

AB-BCB-MI polymers. For example, increased flexibility of the polymer repeat unit may enhance the ability to form crazes [25] and thus lead to improved fracture toughness. Note that AB-BCB-MI-1, which has an additional oxy-1,3-phenylene spacer compared to AB-BCB-MI-2, has a much lower T_g. The greater flexibility corresponding to this lower T_g may contribute to the much more extensive crazing and hence higher fracture toughness for AB-BCB-MI-1 as compared to AB-BCB-MI-2A and 2B. This flexibility argument is invoked as a possible explanation for this difference in crazing behavior, since rubbery modulus data suggest that AB-BCB-MI-1 has a higher crosslink density

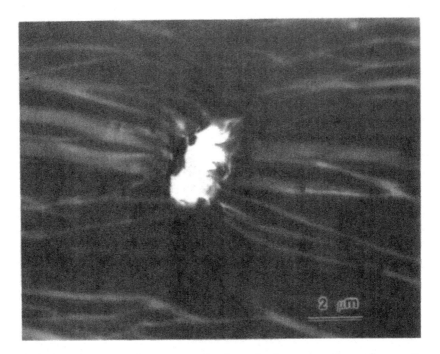

Figure 10 A TEM micrograph taken farther away from the crack tip but still inside the damage zone. Crazes are nucleated and propagate from some sort of inclusion or impurity particle, which has been pulled out during the TEM thin-sectioning process. The crack propagates from right to left.

than AB-BCB-MI-2A, which would be expected to lead to less crazing, not more, in the absence of some other contributing factor.

The crazes observed in Figs. 11 and 12 are neither as distinctive nor as convincing as those shown in Figs. 8 to 10. That the white lines in Figs. 11 and 12 are crazes is based on the argument that the lighter line images must arise from lower mass density compared to the surrounding matrix. Crazing can give rise to lower mass density, whereas shear deformation cannot [6,22,24]. Keep in mind that these TEM micro-

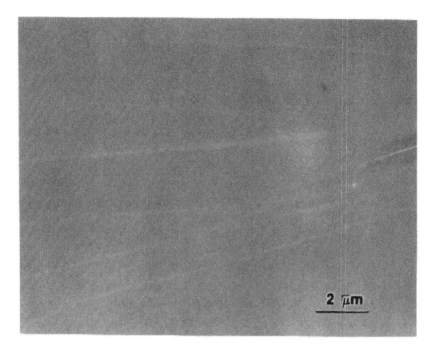

Figure 11 A TEM micrograph taken at the crack tip (see arrow) of the DN-4PB damage zone of AB-BCB-MI-2B. Big, but relatively scare crazes are observed around the crack tip. The crack propagates from right to left.

graphs were of thin sections microtomed from the plane strain damage zone of the DN-4PB specimens; that is, the thin sections were of uniform thickness.

Comparison with other previously investigated systems further suggests that the formation of crazes in thermosets cannot be totally accounted for merely by the crosslink density (i.e., the average molecular weight between crosslinks). For example, AB-BCB-MI-2B appears to have a cross-link density very similar to that of the crosslinkable epoxy thermoplastics (CET) we previously studied [26]. However, in contrast to the crazing of AB-BCB-MI-2B, the fracture mechanism for

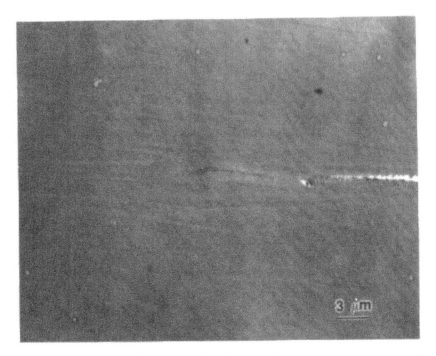

Figure 12   A TEM micrograph taken at the crack tip (see arrow) of the DN-4PB damage zone of AB-BCB-MI-2A. Widespread crazes form around the crack tip region. The crack propagates from right to left.

CET is highly localized crack tip shear yielding of the matrix, with no sign of crazing.

IV. SUMMARY

The present study clearly indicates that crazing can occur in high T_g (> 200°C), high-performance thermosets. However, it is not clear at this stage of our investigation why crazing, instead of shear banding, is the dominant toughening mechanism for these AB-BCB-MI and BCB-DK/HC-ACT polymers.

Although this work has shown how changes in crosslink density and chain flexibility can dramatically affect the extent of crazing, neither the crosslink density nor the flexibility, both of which would appear to be fairly typical for thermosets, would seem to explain why crazing is dominant. The answer as to why crazing is, in fact, the dominant toughening mechanism may lie in other factors not addressed explicitly in the current study, for example, network heterogeneity or some characteristic of the fully cured polymers that follows from the generic chemical structure of these AB-BCB-MI and BCB-DK/HC-ACT polymers.

ACKNOWLEDGMENTS

The authors would like to thank E. I. Garcia-Meitin, R. A. Kirchhoff, K. J. Bruza N. A. Orchard, C. E. Allen, B. T. Allen, E. P. Woo, Dan Sheck, and L. J. Laursen for their discussions and assistance with this work.

REFERENCES

1. C. B. Bucknall, *Toughened Plastics*, Applied Science Publishers, London, 1977.
2. R. P. Kambour, *J. Polym. Sci., Macromol Rev* 7:1 (1973).
3. B. D. Lauterwasser and E. J. Kramer, *Philos. Mag.* 39:469 (1979).
4. A. J. Kinloch, Mechanics and Mechanisms of Fracture of Thermosetting Epoxy Polymers. *Advances in Polymer Science*, Vol. 72 (K. Dusek, ed.), Springer-Verlag, Berlin, 1986, p. 45.
5. A. F. Yee and R. A. Pearson, *J. Mater. Sci.* 21:2462 (1986).
6. H.-J. Sue, *J. Mater. Sci.* 27:3098 (1992).
7. J. L. Bertram, L. L. Walker, J. R. Berman, and J. A. Clarke, U.S. Patent 4,594,291, 1986.
8. K. C. Dewhirst, U.S. Patent 4,786,668, 1988.

9. R. A. Kirchhoff and K. J. Bruza, Benzocyclobutene in Polymer Synthesis, *Prog. Polym. Sci. 18*:85 (1993).

10. D. E. Henton, D. M. Pickelman, C. B. Arends, and V. E. Meyer, U.S. Patent 4,778,851, 1988.

11. D. K. Hoffman and C. B. Arends, U.S. Patent 4,708,996, 1987.

12. P. C. Yang and D. M. Pickelman, U.S. Patent 4,789,414, 1990.

13. P. C. Yang and D. M. Pickelman, U.S. Patent 5,079,294, 1992.

14. H.-J. Sue, E. I. Garcia-Meitin, B. L. Burton, and C. C. Garrison, *J. Polym. Sci. Polym. Phys. Ed. 29*:1623 (1991).

15. H.-J. Sue, R. A. Pearson, D. S. Parker, J. Huang, and A. F. Yee, *ACS Polym. Preprints 29*:147 (1988).

16. H.-J. Sue, E. I. Garcia-Meitin, and D. M. Pickelman, Toughening Concepts in High Performance Epoxies, *Elastomer Technology Handbook* (N. P. Cheremisinoff, ed.), CRC Press, Boca Raton, FL. in press.

17. P. C. Yang, E. P. Woo, M. T. Bishop, D. M. Pickelman and H.-J. Sue, *ACS PMSE 63*:315 (1990).

18. R. A. Kirchhoff, U.S. Patent 4,826,997, 1989.

19. L. S. Tan and F. E. Arnold, U.S. Patent 4,916,235, 1990.

20. R. A. Kirchhoff, U.S. Patent 4,965,329, 1990.

21. M. T. Bishop, K. J. Bruza, S. A. Laman, W. M. Lee, and E. P. Woo, *ACS Polym. Preprints 33*:362 (1992).

22. H.-J. Sue, *Polym. Eng. Sci. 31*:275 (1991).

23. R. A. Pearson and A. F. Yee, *J. Mater. Sci. 26*:3828 (1991).

24. C. S. Henkee and E. J. Kramer, *J. Polym. Sci. Polym. Phys. Ed. 22*:721 (1984).

25. A. S. Argon, Proceedings of the 7th International Conference on Fracture, Houston, 4,2661 (1989).

26. H.-J. Sue and J. L. Bertram, *Colloid Polym. Sci.* (1993).

5
Fracture Behavior of Rubber-Modified High-Performance Epoxies

HUNG-JUE SUE* and E. I. GARCIA-
MEITIN   The Dow Chemical Company,
Freeport, Texas

DALE M. PICKELMAN†   The Dow Chemical
Company, Midland, Michigan

I. INTRODUCTION

Having versatility in their chemical forms as well as in their corresponding impressive physical and mechanical properties, epoxy resins have been, and will continue to be, among the mainstream materials for coatings, adhesives, electrical laminates, and structural components. Depending on specific needs for certain physical and mechanical properties, a combination

---

Current affiliation: Texas A&M University, College Station, Texas.
†Retired.

or combinations of choices of epoxy resin chemistries and/or curing agents can usually be formulated to meet the market demands. However, when the application is structural (i.e., for load-bearing purposes), epoxy resins are unfortunately not different from other engineering polymers; they are either brittle, or notch sensitive, or both. As a result, tremendous effort has been focused on toughness improvement of epoxy resins.

The term *toughness,* in a broader sense, is a measure of material's resistance to failure. Depending on the application and performance of researchers, toughness is usually measured as either the stress or the energy required to fail a specimen under a specific loading condition. More specifically, toughness can be defined as either (1) the tensile strength, (2) the area under the tensile stress–strain curve, (3) the Izod impact strength, (4) the Charpy impact strength, (5) the plane strain critical strain energy release rate (G_{1C}), or (6) the plane strain critical stress intensity factor (K_{1C}). Although the definition of toughness varies from application to application and from researcher to researcher, it is generally agreed that when applications are of safety concerns, the "worst case scenario" toughness definition, that is, either K_{1C} or G_{1C}, should be considered. Since the present work focuses on toughening high-performance epoxies for both automotive and aerospace applications, K_{1C} or G_{1C} (or J_{1C}, if the material is extremely ductile) will be used to determine the toughness of modified epoxy systems in the context of this chapter.

The commonly known approaches for toughening brittle epoxies, which typically have G_{1C} values of less than 200 J/m², include (1) chemical modification of a given rigid epoxy backbone to a more flexible backbone structure, (2) increase of epoxy monomer molecular weight, (3) lowering of the crosslink density of the cured resin via mixtures of high monomer molecular weight epoxies or use of low functionality curing agents, and (4) incorporation of dispersed toughener phase(s) in the cured epoxy matrix. Among these approaches, toughening via dispersed toughener phase(s) has been shown to be most effective and can provide an order-of-magnitude toughness improvement. However, only the highly toughenable epoxies modified with rubber tougheners are known to produce such an impressive toughening effect. These highly toughenable ductile epoxies usually exhibit rather low glass transition temperatures (T_g) and/or low crosslink densities. Consequently, they are not suitable for high-performance structural applications. Thus, research effort is mainly focused on toughness improvement of highly crosslinked high-performance epoxies, where toughening effects have thus far been less effective.

For high-performance, high-T_g brittle epoxies, the toughening effect via rubber modification is usually only incremental. The cause for such a disappointing result is largely attributed to the high crosslink densities of epoxies, which greatly reduces the local molecular mobility. As a result, crack deflection, crack bifurcation, crack pinning, and crack bridging types of energy absorption mechanisms (all relatively low in energy absorption) are among the dominant toughening mechanisms. Nonetheless, the recent work conducted by Kinlock et al. [1] and Glad [2] demonstrates that the tightly crosslinked high performance epoxies can undergo strain softening and strain hardening when tested under compression. These findings indicate that the high crosslink density brittle epoxies can undergo shear yielding–banding as long as the stress state favors such mechanisms. It is therefore conceptually possible to toughen high crosslink density epoxies via shear yield–banding mechanisms, so long as the toughener phase can effectively alter the crack tip stress state from one that favors brittle fracture to one that promotes shear yielding.

The present chapter provides a comprehensive, up-to-date understanding of how to toughen epoxy resins via rubber modification. This includes a brief review of recent developments in epoxy toughening, a brief overview of the use of effective experimental and numerical tools for studying the fracture behavior of epoxy systems, illustrations of important toughening mechanisms observed in rubber-toughened epoxies, and the mechanics and physics relating to the observed toughening mechanisms. Existing concepts and theories in epoxy toughening are reviewed. The detailed procedure for producing these toughened epoxy systems is then discussed. Our intent is to convey the recent discoveries and technology in epoxy toughening; readers should not be limited by the scope of this work in developing advanced toughened epoxy systems for both existing and new applications.

The toughening principles discussed in this chapter are derived from fundamental materials science and mechanics. Therefore, even though operating conditions for effective toughening may vary from polymer to polymer, the toughening principles should be universally applicable to other thermosetting and thermoplastic polymers, and even to adhesive and composite materials if care is taken.

II. TOUGHENING PRINCIPLES

In order to toughen epoxies effectively, the fundamental physics of toughening must be understood; that is, we must know which circumstances promote desirable toughening mechanisms and optimize the toughening effect. Thus, it is essential that we know all the possible operative toughening mechanisms and their relative effectiveness in toughened epoxies, as well as the toughening theories behind the mechanisms. Sultan and McGarry [3] were the first to utilize the concept of rubber toughening in an epoxy matrix. In their study they used carboxyl-terminated acrylonitrile (CTBN) liquid rubber to toughen Epon 828 diglycidyl ether of bisphenol A (DGEBA) epoxy resin. They attributed the observed toughening effect mainly to the crazing of the epoxy matrix, based on the strong dependence of the toughening effect on the rubber particle size and the higher pressure sensitivity under biaxial tension of the large rubber-modified epoxy, compared to those of the small rubber-modified and unmodified epoxies. As a result, rubber particle size and interfacial adhesion between the rubber particles and the epoxy matrix are thought to be critical for promoting crazing in toughened epoxies. Nevertheless,

Bascom et al. [4] accredited the high toughness value of CTBN-modified epoxy to an increase in the plastic zone size. The interpretation was that the triaxial stress associated with the crack tip caused the cavitation of the rubber particles. These cavitated rubber particles induced plastic flow, which is now recognized as massive shear banding, of the matrix around the particles.

Rubber stretching and tearing, also known as the rubber bridging mechanism, are proposed by Kunz and co-workers [5,6] as a major toughening mechanism for rubber-modified plastics. The phenomena are supported by some experimental observations [5–7]. The rubber particle bridging theory proposed by Kunz et al. predicts that about a twofold improvement in toughening can be achieved. However, more recent experimental investigations conducted by others [7–12] indicate that the toughening effect due to rubber stretching and tearing alone cannot account for an order of magnitude increase in toughness improvement. Rather, it is the shear yielding of the matrix that should account for the impressive toughness improvement [13].

A more recent and plausible elastomer toughening concept that uses mechanics to describe the toughening mechanisms of rubber-toughened polymers is proposed by Yee and Pearson [8,9]. In studying the toughening mechanisms of CTBN rubber-modified epoxy systems, they attributed an order of magnitude increase in toughness to the cavitation of the rubber particles, followed by large-scale shear yielding of the epoxy matrix. They emphasized the importance of the sequence of toughening events.

In an attempt to further support the preceding toughening concept, Yee and co-workers [7,9,14–17] conducted a series of experiments and clearly demonstrated that for significant plastic shear banding to operate under constrained conditions in both thermoplastic and thermoset systems, cavitation of the toughener phase is essential, via internal rubber particle cavitation, debonding at the interface, or crazing mechanisms. In other words, there is a sequence of toughening events (i.e., cavitation occurs first, followed by shear banding) and a causal relationship (i.e., without the cavitational process the shear banding mechanism cannot take place) involved in the toughening process.

III. EXPERIMENTAL WORK

In order to study the toughening mechanisms and the sequence of failure events in toughened epoxies, it is important that one microscopically investigate and mechanically understand exactly how the crack is evolved during the failure process. To do so, an effective mechanical testing method has to be used to create and preserve the crack evolution process, followed by an achievable microscopy technique to observe the failure events. Furthermore, fracture mechanics tools relating to the crack evolution events are required to gain fundamental knowledge of how epoxies can be toughened. These experimental tools are introduced and discussed below.

A. Materials

Throughout this work D.E.R.* 332 (DGEBA) epoxy resin cured stoichiometrically with 4,4'-diaminodiphenylsulfone (DDS) is used to model high-crosslink-density epoxy matrices.

The rubber modifiers used in this study include core–shell butadiene rubbers (CSR) [18] with various shell compositions (Table I) and dispersed acrylic rubber (DAR) [19,20]. A curing schedule of 180°C for 2 hours and 220°C for 2 hours was used for both toughened and untoughened epoxies.

B. Damage Creation

After the 0.635-cm sample plaques were completely cured, they were slowly cooled to room temperature inside the oven.

---

*Trademark of The Dow Chemical Company.

Single-edge-notch three-point-bend (SEN-3PB) specimens, having dimensions of $6.35 \times 1.27 \times 0.635$ cm were used for K_{1C} measurements. Since the damage evolution process ahead of the crack tip is obliterated by the SEN-3PB test, information concerning the sequence of failure events and the role(s) the rubber particles play in the toughening process cannot be definitively obtained. Also, knowledge concerning the possible interactions among the operative toughening mechanism cannot be gained. Therefore, the double-notch four-point-bend (DN-4PB) method [21,22] is utilized to generate and preserve the crack evolution process (Fig. 1).

In this work, the DN-4PB specimens, having dimensions of $12.7 \times 1.27 \times 0.635$ cm were used to generate a subcritically propagated crack for the investigation of the crack tip damage evolution process of rubber-toughened epoxies.

C. Microscopy

To understand the toughening mechanism(s) and the sequence of failure events in rubber-modified epoxies, it is imperative that the damage zone of the modified systems be studied. Ease of use and straightforward sample preparation procedures make reflected optical microscopy (OM) and scanning electron microscopy (SEM) among the most utilized microscopy techniques for studying both morphology and failure mechanisms of all materials. These techniques are especially useful for determining any surface features that exhibit topographical variations due to the fracture surface failure characteristics and phase morphology of modified systems. These microscopic techniques, however, only provide information relating to the fracture surface, such as the crack path, the rubber particle size, shear lip, plastic drawing of the material, remnants of crazes (if any), crack pinning, basic longitudinal texture, mackerel pattern, and so on. When information concerning the subfracture surface zone (SFSZ) is of interest, investigation of the SFSZ of the damaged specimen using transmitted OM (TOM) and/or transmission electron microscopy (TEM)

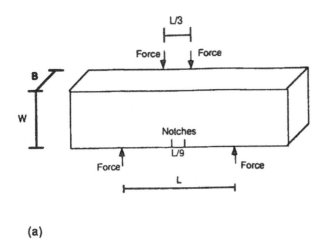

(a)

Shear Yielding Zone

Fracture Surface

Sub-Fracture Surface
Damage Zone

(b)

Figure 1   (a) Schematic of the DN-4PB geometry. (b) Schematic of
the regions from the DN-4PB specimen for scanning electron microscopy (SEM), reflected optical microscopy (OM), and transmission
electron microscopy (TEM) investigations.

becomes necessary. Also, damage features relating to the SFSZ, which usually account for most of the fracture energy dissipation in polymers, cannot be obtained from fracture surface studies. As a result, the SFSZ investigation is essential for complete understanding of the contributing toughening mechanisms in rubber-modified epoxies. The detailed sample preparation procedures for OM and TEM investigations can be found elsewhere [23–26].

D. Mechanics Tools

1. FRACTURE TOUGHNESS MEASUREMENTS

Polymeric materials are, as pointed out earlier, viscoelastic and viscoplastic. Strictly speaking, the conventional linear elastic fracture mechanics (LEFM) approach [27] cannot be applied to characterize polymer fracture behavior. Nevertheless, when the crack tip plastic zone is small, compared to the width and thickness of the specimen, and the testing rate is significantly higher than the characteristic polymer relaxation time constant, the LEFM approach may still hold. Care must be taken to assure valid fracture toughness measurements.

To measure a valid K_{1C} of the rubber-modified epoxies, a mechanical testing procedure following ASTM D399-83 is conducted in this study. A Sintech screw-driven mechanical tester, with a crosshead speed of 0.0508 cm/min, is used throughout the work.

2. STRESS ANALYSIS

Toughening phenomena in rubber-modified epoxies result, per se, from the differences in mechanical response of each component (i.e., the rubber inclusion or the epoxy matrix) to the external mechanical disturbance, such as tensile load or impact load. Furthermore, since all polymers are viscoelastic and viscoplastic, finite time is required to build up the stress level for each component to undergo irreversible deformation. Therefore, adequate mechanics tools are needed for definitive descriptions of the role(s) the rubber particles play and the sequence of toughening events in the polymer-toughening process.

The conventional LEFM and elastic plastic fracture mechanics (EPFM) approaches [27,28] directly characterize the fracture behavior of toughened polymers using basic material parameters, such as Young's modulus and Poisson's ratio. However, these parameters are sometimes inappropriate in connection with the operative micromechanisms in the damage zone.

Micromechanics, using both stress and energy approaches [29,30], was developed to link the micromechanical behavior of polymers to their global mechanical behavior. Unfortunately, the lack of knowledge concerning the fundamental physics of how and why a micromechanism is triggered makes it impossible to utilize fracture mechanics tools directly to predict the toughening event(s) upon fracture. Hence, alternative approaches, such as classical continuum mechanics and finite element methods (FEM) [31], are implemented and found to be extremely useful as supplemental tools for qualitative or semiquantitative modeling of the micromechanical behavior of polymers [32–34]. In other words, only the well-established mechanics tools (such as slip–line field theory [35,36], the Irwin [37], Goodier [38], Dewey [39], and Eshelby [40] equations, and FEM) are used to study both the crack tip stress field and the stress state under which a toughening mechanism can be triggered. Applications of combinations of these approaches can greatly improve quality in the design of toughened polymers [32–34].

In studying the stress disturbance due to the mismatch of elastic constants between the matrix and the toughener phase, the well-established classical equations derived by Goodier, Dewey, Good, and Eshelby can all be used to describe explicitly the linear elastic stress and strain fields around the inclusion phase, so long as the inclusion is either spherical or ellipsoidal. In order to estimate the stress and strain fields of an inclusion in front of the crack tip, superpositions of Irwin's equation, slip-line field theory, and one of the preceding classical equations (e.g., Dewey's equation), as a first approximation, can help describe how the toughener phase responds to the crack perturbation [34]. This approach has been shown to provide a plausible interpretation of why cavitation of the rubber particles is crucial in promoting the shear yielding–banding mechanism in polymer toughening.

When the material is nonlinear and/or the morphology of the toughened system is too complex, the FEM is found to be extremely powerful in elucidating the unusual postyielding behavior of polymers [32–34]. However, care has to be taken when the nonlinear behavior of polymers is incorporated into the FEM simulation process. Experimental verification of simulation results is recommended because of oversimplification due to the FEM analysis.

IV. STUDIES ON TOUGHENING MECHANISMS

In an effort to shed some light on how to toughen high-performance epoxies more effectively, research on the influence of types of rubber, particle size, and particle dispersion characteristics in toughening is being conducted. These efforts, along with other relevant work, it is hoped, will result in a more definitive concept of toughening.

As shown in Tables 2 and 3, it is clear that the type of rubber, particle size, and particle dispersion characteristics do affect toughening in brittle epoxies. In order to determine the exact role(s) the rubber particles play and investigate the operative toughening mechanisms in the toughening process, the DN-4PB damage zone of each brittle epoxy system is stud-

Table 1 Compositions of the Core–Shell Rubber Particles

Rubber particle	Core/ shell	Particle diameter (nm)		Shell composition				
		Experimental	Theoretical	S	MMA	AN	GMA	δ[d]
Core	100/0	119	—	—	—	—	—	—
CSR-A	84/16	Not measured	126	3.6	3.6	4.0	4.8	10.3
CSR-B	84/16	127	126	4.8	4.8	4.0	2.4	10.2
CSR-C	84/16	127	126	6.0	6.0	4.0	0.0	10.1
CSR-D	84/16	Not measured	126	7.0	7.0	2.0	0.0	9.7
CSR-E	84/16	Not measured	126	8.0	8.0	0.0	0.0	9.3
CSR-F	75/25	133	131	7.5	7.5	6.25	3.75	10.2
CSR-G	65/35	144	140	10.5	10.5	8.75	5.25	10.2
Neat epoxy								10.3

[a]Core composition, 7% styrene and 93% butadiene.
[b]Brice-Phoenix Universal light scattering photometer.
[c]Parts of shell: S = styrene. MMA = methyl methacrylate, AN = acrylonitrile, GMA = glycidyl methacrylate.
[d]Estimated via group energy methods [see, eg., P. A. Small, J. Appl. Chem. 3:71 (1953)].

ied using TOM and TEM. Only on rare occasions when the fracture surface accounts for most of the roughening effect will SEM be utilized.

A. Core–Shell Rubber Modification

In this study, 10 wt % of CSR-B particles (Table 1), having a uniform particle size of approximately 0.12 μm and a T_g of about $-80°C$, was incorporated into DGEBA epoxy resin to study the effectiveness of small rubber particles in toughening brittle epoxies. This system exhibits random dispersion and greater toughening effect than those using DAR or CTBN rubber modification [9] (Table 2). It is therefore important to discover the role of the CSR particles in the toughening process.

Since the size of the CSR-B particles is too small for TOM observation, only TEM is utilized. The TEM micrographs of the DN-4PB crack tip damage zone, as shown in Figure 2, indicate that the small, 0.12-μm CSR-B particles (see Table 1) are quite effective in promoting extended shear yielding of the brittle epoxy matrix. The matrix has undergone approximately 60% plastic deformation based on the fact that the

Table 2 Summary of Fracture Toughness of Various Rubber-Modified Epoxy Systems

Material	K_{1C} (MPa m$^{1/2}$)	G_{1C} (J/m^2)	T_g[a] (°C)
DGEBA epoxy/DDS	0.83 ± 0.03	180	220
DGEBA epoxy/DDS/ 10% CSR-B rubber	1.20 ± 0.04	490	219
DGEBA epoxy/DDS/ 10% DAR rubber	1.08 ± 0.06	422	193
DGEBA epoxy/DDS/ 10% CTBN rubber [9]	0.91 ± 0.05	242	220[b]

[a]Second-heat midpoint T_g value is reported.
[b]Post-cured at 250°C.

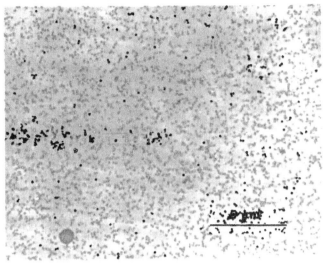

(a)

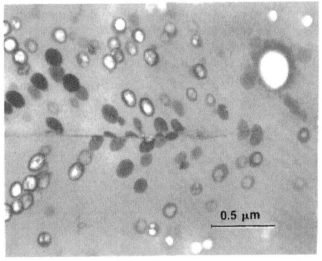

0.5 μm

(b)

Figure 2 TEM micrographs taken at the crack tip of the damaged DN-4PB specimen of CSR-B–modified epoxy system at (a) low magnification and (b) high magnification. A 60% elongation of the CSR rubber particles at the crack tip is found, which indicates that the brittle epoxy is capable of undergoing extended plastic deformation. The crack propagates from left to right.

144

rubber particles at the crack tip elongated by as much as 60%, compared to the undeformed spherical rubber particles (Fig. 3). The major toughening mechanisms in this system appear to be cavitation of the CSR-B particles, followed by shear yielding of the matrix. In other words, the cavitation of the rubber particles helped relieve the plane strain constraint induced by the thick specimen and the sharp crack; the octahedral shear stress component around the crack tip is, as a result, greatly raised and causes extended yielding of the matrix. For clarity, the schematic of the toughening sequence of events is drawn in Fig. 4.

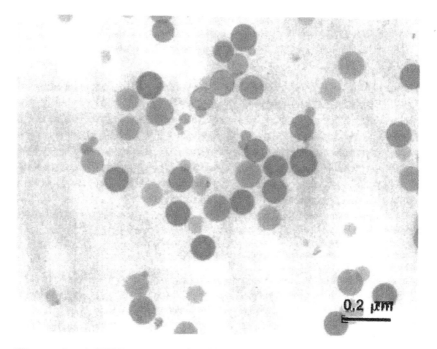

Figure 3 A TEM micrograph taken at the undamaged region of the CSR-B–modified epoxy system. The CSR rubber particles appear to be spherical. No cavities are found inside the rubber particles.

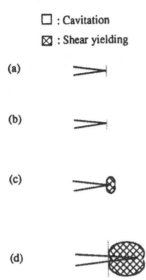

Figure 4 A sketched sequence of the toughening mechanisms of the CSR-B–modified epoxy system. (a) The initial starter crack: (b) formation of a cavitation zone in front of the crack tip when the specimen is initially loaded: (c) formation of initial shear-yielded plastic zone around the crack tip when the hydrostatic tension is relieved by the cavitation of the rubber particles. (d) Once the buildup of shear strain energy reaches a critical value, the material begins to undergo shear yielding, and the crack propagates, leaving a damage zone surrounding the propagating crack before the crack grows unstably. The plastic zone is not drawn to scale.

Interestingly, the rubber particle at the crack tip (Fig. 2b) appears to be partially broken. One can thus say intuitively that prior to the crack advancement, the rubber particles in front of the crack tip first cavitate and induce extended shear yielding of the matrix by as much as 60% plastic strain. After cavitation and crack advancement, the elongation of the rubber particles can reach as much as several hundred percent;

crack bridging due to the rubber particles is then possible. However, the rubber particles were cavitated before particle bridging took place. The rubber particles were also partially elongated prior to particle bridging. As a result, the toughening effect due to particle bridging is probably minimal, if it occurs at all.

In addition to the rubber particle cavitation/matrix shear yielding and the possible rubber bridging mechanisms, crack bifurcation and crack deflection mechanisms are also observed (Figs. 5 and 6). These mechanisms are supplemental to the toughening of the brittle epoxy matrix. Although the

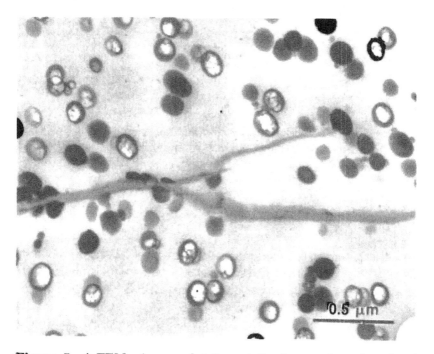

Figure 5 A TEM micrograph taken at the damaged crack wake if the CSR-B–modified epoxy system. Crack bifurcation is observed. The crack propagates from left to right.

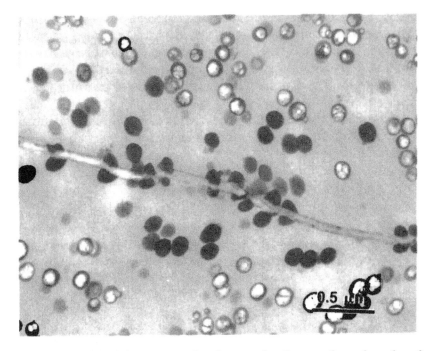

Figure 6 A TEM micrograph taken at the damaged crack wake of the CSR-B–modified epoxy system. The crack propagates through the rubber particles, instead of propagating around the rubber particles. The rubber particles appear to have deflected the crack path. The crack propagates from left to right.

crack appears to follow a path through the CSR-B particles in front of the crack tip, the degree of crack deflection is minimal. This is because the size of the CSR-B particles is too small to deflect the crack effectively.

It is noted that some of the CSR-B rubber particles in the wake and at the crack tip do not show cavitation. This abnormality is mainly due to the OsO_4 staining solution, which tends to penetrate through the crack path faster (because of capillary action) than through the matrix. As a result, the rubber particles adjacent to the crack face are overstained, and the cavities of these rubber particles are covered by the stain. This has also been observed and verified in other systems [24,41].

B.   Dispersed Acrylic Rubber Modification

Having exactly the same curing schedule and weight percent of rubber, the DAR-modified brittle epoxy system exhibits a lower toughening effect than that of the CSR-modified epoxy system (Table 2). The DAR particle has an average size of about 0.4 μm and a T_g of about $-60°C$, values different from those of the CSR particles.

When the DAR is used to toughen the brittle epoxy matrix, the response of the DAR particles to the advancing crack appears to be quite different from that of the CSR-B particles. Except for the particles that are in contact with the crack planes, the DAR particles stay spherical around the damaged crack tip (Fig. 7). In other words, no sign of extended shear yielding of the matrix is found. This is probably the reason why the DAR modification is less effective than the CSR-B modification in toughening brittle epoxies.

The major toughening mechanisms in DAR-modified epoxy are found to be crack deflection and crack bifurcation (Figs. 8 and 9). Since the DAR particle size (≈ 0.4 μm) is bigger than that of the CSR particles (0.12 μm), the crack deflection mechanism is more effective in the DAR-modified epoxy than in the CSR-modified system. Interestingly, voids in the epoxy matrix can also help promote the crack deflection mechanism (Fig. 10). Therefore, it is conceivable that well-dispersed voids with appropriate sizes (preferably a few times bigger than the characteristic crack tip radius) are as effective as the rubber particles in deflecting the crack. Crack bifurcation appears to occur in the DAR-modified epoxy (see big arrows in Fig. 9), as well. Microcracking induced by the DAR particles also seems to oper-

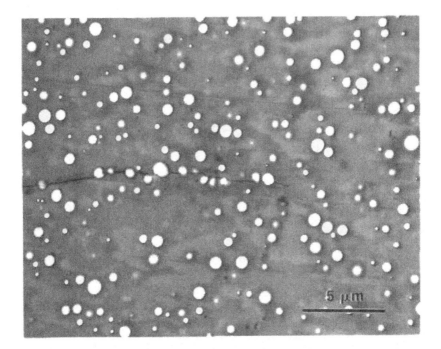

Figure 7 A TEM, micrograph taken around the crack tip of the damaged DN-4PB specimen of the DAR-modified epoxy. Only the particles that contacted the crack faces are deformed. The adjacent rubber particles appear to be unaffected by the propagated crack. The crack propagates from left to right.

ate in this system (small arrow in Fig. 9). In general, however, the crack deflection, crack bifurcation, and microcracking toughening mechanisms are less effective than the shear yielding mechanism in toughening epoxies.

Owing to the fact that the DAR particles cannot be effectively stained via any existing staining technique, a novel selective solvent sorption and staining technique [24] was developed to stain the crack tip damage zone of the DAR-modified epoxy system. As indicated in Fig. 11, the DAR

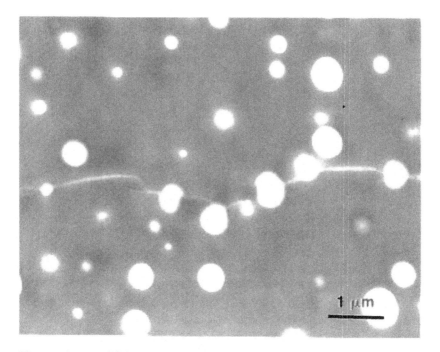

Figure 8 A TEM micrograph taken at the damaged crack wake of the DAR-modified epoxy. The crack is deflected by the DAR particle in front of it. The crack propagates from left to right.

particles are, in fact, cavitated around the crack tip and the crack wake. This finding implies that the cavitation of the rubber particles alone is not sufficient to cause extended shear yielding of the matrix. The stress state at which the rubber particles cavitate should also exceed a certain critical value to activate the yielding of the matrix [42].

It is noted that even with the use of the novel selective solvent sorption and staining technique, no crack/particle bridging mechanism can be observed. Therefore, it is still not certain whether or not crack bridging takes place in this system.

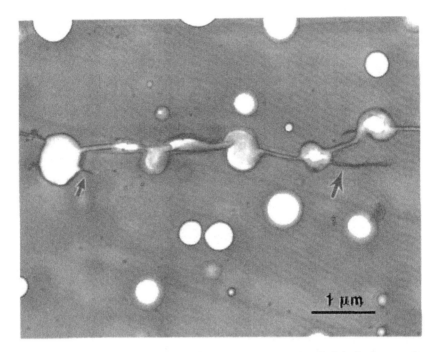

Figure 9 A TEM micrograph taken at a region behind the crack tip of the DAR-modified epoxy. The crack bifurcation mechanism (big arrow) can be easily observed. Microcracking (small arrow) appears to take place as well. In addition, a possible particle/crack bridging mechanism appears to be present. The crack propagates from left to right.

C   Variation of Core–Shell Rubber Interfacial Characteristics

In an attempt to study whether or not the dispersity of rubber particles and the grafting efficiency of rubber particles to the matrix affect the toughening effect, the composition and thickness of the shell on the CSR particle are varied. A total of seven types of CSR particles are investigated. They are listed in Table 1. The fracture toughness as well as the T_g of each of

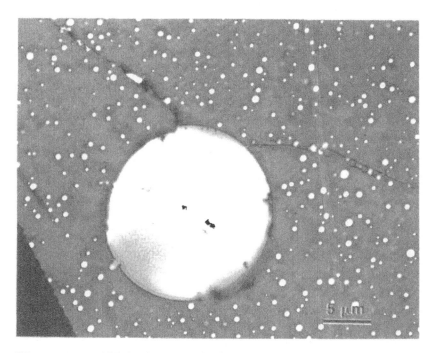

Figure 10 A TEM micrograph taken at the damaged crack wake of the DAR-modified epoxy. The crack is deflected by a void. The crack propagates from left to right.

the CSR-modified high-performance epoxies are shown in Table 3.

1. VARIATION OF GLYCIDYL METHACRYLATE CONTENT IN THE SHELL

Since glycidyl methacrylate (GMA) contains an epoxide functional group, the GMA component in the shell is believed to react chemically with, and therefore graft to, the epoxy matrix. Consequently, the GMA concentration can affect not only the dispersion of the CSR in the matrix via solubility parameter (δ)

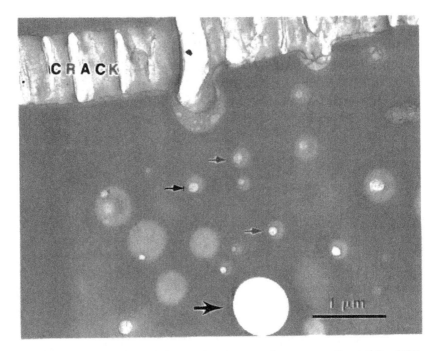

Figure 11   A TEM micrograph taken at the core region of the DN-4PB sample. The thin sections are stained with styrene/OsO4 and treated with periodate anion [24]. It is evident that the DAR particles do cavitate (indicated by small arrows) at the SFSZ. The big arrow indicates a hole formed by a rubber pull-out during the thin-sectioning process. The crack propagates from left to right.

changes (Table 1), but also the efficiency of chemcial grafting to the epoxy matrix. The variation of the GMA content from 30% to 0% by weight in the shell (i.e., CSR-A, CSR-B, and CSR-C) appears to alter the dispersion of the CSR particles in the epoxy matrix somewhat, that is, from multimodal dispersion, to random dispersion, and to globally random but locally clustered dispersion (Fig. 2, 12 and 13). Nevertheless, the toughening effect observed for the three systems is practically the same (Table 3). This implies that the chemical bonding between the

Table 3 Summary of K_{1C}, G_{1C} and T_g of Designed CSR-Modified Epoxy Systems

DER 332 epoxy resin/DDS	K_{1C} (MPa m$^{1/2}$)	T_g (°C)	G_{1C}(J/m^2)
Neat resin	0.83 ± 0.03	220	180
Modified with 10 wt % CSR			
—A	1.20 ± 0.07	222	490
—B	1.20 ± 0.04	219	490
—C	1.30 ± 0.09	223	580
—D	1.07 ± 0.05	229	390
—E	1.05 ± 0.03	224	380
—F	1.33 ± 0.07	223	620
—G	1.37 ± 0.06	220	640

[a]A 63% confidence interval is used.
[b]Second-heat midpoint T_g is reported.
[c]$G_{1C} = K_{1C}^2 (1 - v)/E$.

rubber particles and the epoxy matrix is not critical in toughening by CSR. In other words, the possible molecular intermixing and/or the physical polarity interactions at the interface between the shell of the CSR and the epoxy matrix may be sufficient to maintain the interfacial integrity. This also implies that the cavitational strength of the CSR particle is weaker than that of the resultant interfacial strength. The toughening mechanisms in these systems are quite similar. Except for the CSR-C system where an additional crack deflection mechanism is operative (due to local clustering), the major toughening mechanisms in these three systems are cavitation of the rubber particles, followed by shear yielding of the matrix.

2. VARIATION OF ACRYLONITRILE CONTENT IN THE SHELL

The preceding study on CSR-A–, CSR-B–, and CSR-C– modified epoxies suggests that the GMA content does not

(a)

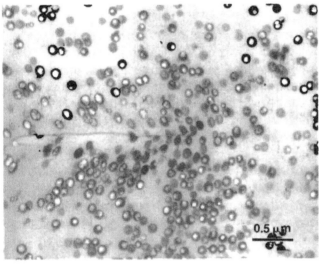

(b)

Figure 12 TEM micrographs of a damaged DN-4PB sample of the CSR-A–modified epoxy system at (a) a lower magnification and (b) a higher magnification. A multimodal dispersion of the CSR particles is observed. The crack propagates from left to right.

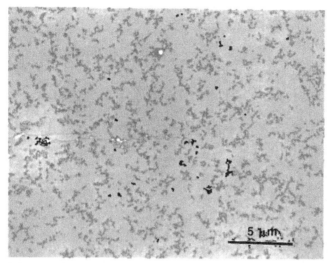

(a)

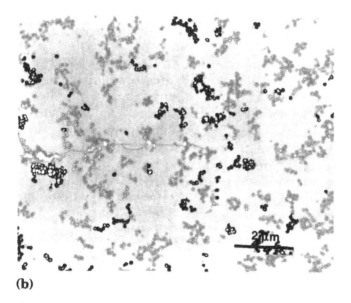

(b)

Figure 13 TEM micrographs of a damaged DN-4PB sample of the CSR-C–modified epoxy system taken at (a) low magnification and (b) high magnification. The CSR particles are globally well dispersed, but locally clustered. The crack propagates from left to right.

157

affect the toughening effect. Consequently, the GMA is omitted hereafter.

When the acrylonitrile (AN) content of the shell is varied from 25% to 0% by weight (i.e., CSR-C, CSR-D, and CSR-E), the δ of the shell varies from about 10.1, to 9.7, to 9.3. Since the δ for the epoxy matrix is $\approx$ 10.3, the dispersion of the CSR particles in the epoxy matrix will likely be changed from good dispersion to poor dispersion. Indeed, as shown in Figs. 13 to 15, the dispersion of the CSR particles is altered from globally random but locally clustered dispersion, to more local particle clustering, to large-scale agglomeration of the CSR particles in the epoxy matrix. The toughening effect is, as anticipated, good for the CSR-C–modified system and poor for both the CSR-D– and CSR-E–modified systems.

The toughening mechanisms in these systems are quite different; rubber particle cavitation, matrix shear yielding, and limited crack deflection are found for the CSR-C–modified system, whereas only the crack deflection mechanism is observed for both the CSR-D– and CSR-E–modified systems.

The investigation of the toughening mechanisms in this series suggests that, qualitatively speaking, a certain degree of rubber particle clustering is desirable. Local clustering of the CSR particles will not affect the toughening effect. However, when the clustering increases in magnitude, the local stress field overlap causes the clustered rubber particles to act like a big, irregularly shaped rubber particle, which functions like the Proteus 5025 particle [23]. As a result, the individual rubber particles can no longer cavitate to relieve the plane strain constraint. This results in brittle failure of the matrix.

3. VARIATION OF SHELL THICKNESS

The thickness of the shell may not only help retain the shape of the CSR particle during part fabrication but also potentially effect how the copolymers arrange themselves in the shell. The possible rearrangement of the copolymers in the

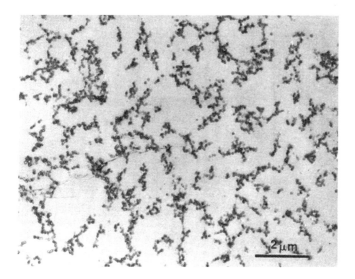

(a)

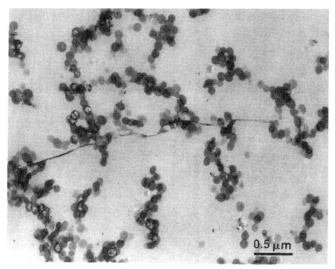

(b)

Figure 14 TEM micrographs of a damaged DN-4PB sample of the CSR-D–modified epoxy system taken at both (a) a low magnification and (b) a high magnification. Cavitation of CSR particles is highly suppressed by the poor dispersion of the CSR particles. The crack propagates from left to right.

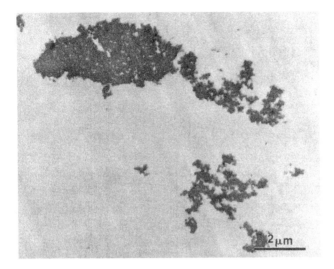

(a)

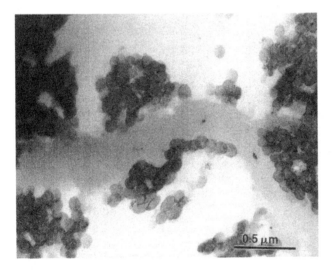

(b)

Figure 15 TEM micrographs of the CSR-E–modified epoxy system. (a) Without glycidal methacrylate and acrylonitrile in the shell, the CSR particles severely agglomerate and form irregular domains. (b) At the damage zone, only crack deflection is observed; rubber particle cavitation is suppressed; the crack propagates from left to right.

160

shell will probably affect the dispersity of the CSR particles in the epoxy matrix. This will, in turn, alter the failure process in the CSR-modified epoxy systems.

When the shell thickness is varied by changing weight percent from 15%, to 25%, to 35% in the shell (i.e., CSR-B, CSR-F, and CSR-G) while all other parameters are kept the same (Table 1), the dispersions of the CSR particles do change from random dispersion (Fig. 2) to globally well dispersed but locally clustered (Figs. 16 and 17). This implies that the copolymers in the shell have somehow rearranged themselves because of variations in shell thickness. Questions concerning how the shell thickness affects copolymers rearrangements still await further investigation.

The toughening effect exhibited by the three systems is quite surprising. Random dispersion of the CSR particles does not give optimal toughening (Table 3). Instead, there appears to be a synergistic toughening effect due to the nonrandom dispersion of the rubber particles. An average of $\sim$ 640 J/m^2 in G_{1C} can be obtained for the CSR-G modified brittle epoxy system. In comparison, the G_{1C} is only $\sim$ 490 J/m^2 in the CSR-B modified system, where the CSR particles disperse randomly. As shown in Fig. 17, it is apparent that the local clustering of the CSR-G particles in the epoxy matrix not only preserves the important rubber cavitation and matrix shear yielding mechanisms but also triggers a vigorous crack deflection mechanism. This additional crack deflection mechanism is probably the main reason for such a synergistic toughening effect.

The study of the effect of rubber particle dispersion in epoxy toughening brings up a valuable concept concerning approaches for toughness optimization. That is, the dispersion of the rubber particles can influence, and even alter, the toughening mechanisms in toughened epoxies. Furthermore, based on this effort, it is clear that if combinations of certain toughening mechanisms coexist during the toughening process, then a synergistic toughening effect may be obtained.

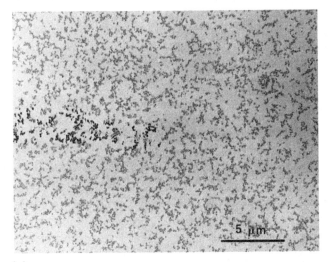

(a)

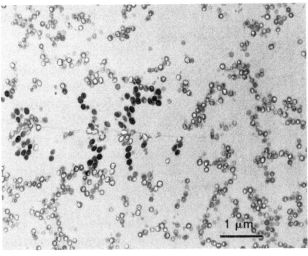

(b)

Figure 16 TEM micrographs of a damaged DN-4PB sample of the CSR-F–modified epoxy system taken at both (a) a low magnification and (b) a high magnification. The CSR particles are globally well dispersed. However, the particles are locally clustered. The crack propagates from left to right.

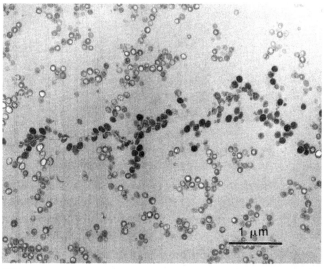

(a)

(b)

Figure 17 TEM micrographs of a damaged DN-4PB sample of the CSR-G–modified epoxy system taken at both (a) a low magnification and (b) a high magnification. The CSR particles are globally well dispersed. However, the particles are locally clustered. The crack propagates from left to right.

V. APPROACH FOR MAKING TOUGHER
EPOXY SYSTEMS

To toughen epoxies, the traditional chemistry-oriented approaches have been to link the rigid epoxy molecules chemically to more flexible chains, physically blend flexible monomers into the rigid monomers, use less reactive curing agents, or to combine the methods. Indeed, these approaches do produce ductile epoxies (for engineering applications) exhibiting better molecular chain flexibility, which can improve the tensile elongation to break from several percent to over 20%. However, the notch sensitivity of the epoxies made via these modifications is usually not much altered. As a result, the brittle epoxies modified through chemical means are still brittle. They are not suitable for structural applications.

An alternative approach, which utilizes the presence of a dispersed inhomogeneity in the epoxy matrix to promote an extended damage zone when fracture occurs, is found to be, in general, far more effective in toughening epoxies. This approach, however, does not always guarantee effective toughening of brittle epoxies. The type of rubber, particle size, and dispersion can all greatly affect toughening. The understanding of the exact role or roles the dispersed phase (usually the rubber particles) plays in the toughening process is still mostly unclear. The present work, which is only part of a larger effort to optimize the toughening effect in epoxy systems, focuses on using materials science and mechanics tools to gain understanding that will make it possible to improve the highly versatile but brittle epoxy matrices. Based on the present experimental work, as well as related work conducted by others, an up-to-date toughening concept is discussed below.

A. Materials Selection

The process of selecting an appropriate epoxy matrix for structural applications usually begins with the requirements of its

T_g, stiffness, and, possibly, its physical and chemical properties. However, owing to its inherent brittleness and/or notch sensitivity, the epoxy matrix needs to be toughened. The physical and chemcial properties of the epoxy matrix, which are material characteristics and cannot be altered, are governed by the chemical species in the molecule. The T_g and stiffness of the epoxy matrix is, however, strongly dependent on the structural arrangement of the molecules. In other words, the rigidity of the molecular chain and the crosslink density of the epoxy may dictate both the T_g and stiffness of the epoxy matrix.

The work conducted by Yee and Pearson [7–9] convincingly demonstrates that the low crosslink density epoxies are far more readily toughened than the high crosslink density epoxies. Combining the experimental observations by Yee and Pearson and the understanding that T_g is governed by the rigidity of the epoxy backbone, Portelli et al. [43] are able to develop high T_g and highly toughenable fluorene epoxies. The fluorene epoxies are synthesized and cured in such a way that the local rigidity of the molecule is enhanced by the fluorene backbone structure, while allowing the molecular weights between the crosslinks to be high. With only 7.5% of CSR modification (obtained from Rohm & Haas), an improvement of over 300% in G_{IC} can be obtained without sacrificing the matrix T_g [43]. Therefore, in selecting an epoxy matrix for structural applications, the preferred choices are low crosslink density and high local rigidity epoxy molecules.

In selecting rubber tougheners, it appears that, depending on the application, the type of rubber, size, and dispersion of the rubber particles all play important roles in toughening of brittle epoxies. The use of soluble rubbers to toughen epoxies can result in (1) lowering of the T_g of the epoxy matrix, (2) difficulty in morphology control, and (3) poor reproducibility of the product performance [44,45] These shortcomings are highly undesirable in high-performance aerospace applications. Consequently, preformed rubber particles with controllable and reproducible particle size and shape have recently been developed. These preformed rubber particles include the DAR and CSR particles investigated in this work.

Investigation of the CSR-modified epoxy systems shows that the highly crosslinked brittle epoxy system can undergo shear yielding around the propagated crack tip, even when the crack experiences the plane strain mode-I loading condition. The major toughening mechanisms for these systems are found to be the cavitation of the rubber particles, followed by the formation of a shear-yielded zone around the propagated crack. If this rather small-scale yielding can be further extended, then, potentially, an order of magnitude increase in fracture toughness can be achieved. Nonetheless, not all the rubbers can produce the same result as that of the CSR rubber particles. The CTBN rubber used by Yee and Pearson [9] and the DAR particles studied in this work all show negative results in inducing the extended shear yielding mechanism. Even with the use of CSR particles, if the particle dispersion is poor (as in the CSR-D and CSR-E systems discussed in this study), the rubber particle cavitation and matrix shear yielding mechanisms are suppressed. Thus, the type of rubber, rubber particle size, and particle dispersion in the matrix will induce entirely different toughening mechanisms in modified brittle epoxies.

The degree of interfacial adhesion between the toughener phase and the matrix is also critical in toughening. The strength of the adhesion necessary at the interface to both relieve the triaxial tension and trigger localized shear banding is still not clear. The present study and previous works on polyethylene-modified polycarbonate [45] and rubber-modified nylon [47] do, however, suggest that chemical bonding at the interface is not essential to assure sufficient interfacial bonding. Molecular interlocking, physical interactions, and thermal stress effect can all either enhance or deteriorate the interfacial adhesion between the toughener phase and the matrix. In general, it is believed that an intermediate interfacial bonding (i.e., neither too strong nor too weak an interfacial strength) should be beneficial for effective toughening. This conjecture, nevertheless, cannot be verified until quantitative measurement of interfacial strength can be established.

B. Promoting Desirable Toughening Mechanisms

The toughening principles for relatively ductile polymers have been reviewed and discussed by Yee and Pearson [7-9] and Kinloch [10]. In brittle epoxies, Garg and Mai [48] have summarized the important mechanisms for toughening. Based on the experimental work conducted in the present study, an approach for toughening high performance epoxies is discussed below.

The present work shows that shear yielding can occur in brittle epoxy systems. It also shows that shear yielding and crack/particle bridging mechanisms cannot both be dominant toughening mechanisms in uniform particle size rubber-modified systems [15-18,23] At this stage, however, it is still not clear whether an order-of-magnitude increase in fracture toughness may be attained in brittle epoxy systems. In order to clarify this uncertainty, it is necessary that the nature of shear yielding and shear banding in brittle epoxy systems as well as the quantitative estimation of the crack tip stress field before and after the rubber particles cavitate be understood.

To test whether or not a synergistic toughening effect due to both shear yielding and crack/particle bridging exists, a bimodal size distribution of rubber particles has been examined by Pearson and Yee [7]. They manage to incorporate both the 1 to 2 μm and 10 to 20 μm CTBN rubber particles in D.E.R. 331 epoxy resin and cure with piperidine. They conclude that the bimodal rubber particle modification does not further improve the fracture toughness of the system, compared to that of the unimodal (1 to 2 μm particle size) distribution system.

More recently, a new toughening mechanism, termed the *croiding* mechanism [49], is found to exhibit both high shear plasticity and high dilatational plasticity in the epoxy matrix. This mechanism, although it has only been investigated in low-crosslink-density epoxy systems, may be an alternative route for effective toughening of brittle epoxies. Although there is still much to be done to understand exactly how the croid is formed and the conditions to activate it, toughening epoxies by croiding is shown to be as effective as, if not better than, use of the shear yielding mechanism. An order-of-magnitude improvement in toughness via the croiding mechanism has been reported [50]. It appears that the intrinsic properties of both the matrix and the rubber particles, as well as the rubber particle concentration, play significant roles in causing the formation of croids. It is also shown that the stress state under which croiding occurs must be highly triaxial [49,50]. Since the croiding mechanism can operate in a highly localized manner, it can potentially be utilized for both adhesive and composite toughening applications, where the stress is almost always highly triaxial and the matrix material is very thin.

If, somehow, the shear yielding and plastic dilatation of the matrix cannot be induced for other reasons, alternative mechanisms, such as crack/particle bridging [5,6], microcracking [51], crack deflection [52,53], crack bifurcation [54.55], and crack pinning [56,57], should be considered. These mechanisms appear to occur in the rubber-modified systems quite readily. As pointed out earlier, for the crack bridging mechanism to occur, the rubber particles need to be large with respect to the characteristic crack tip radius. The interfacial adhesion between the rubber particle and the matrix needs to be strong. The rubber particle needs to be drawable as well. Therefore, composite rubber particles, such as the Proteus particles [23] and occluded rubber particles that cannot be extensively stretched, are not suitable for promoting the crack bridging mechanism.

For the microcracking mechanism to occur, the interfacial

adhesion between the matrix and the toughener phase does not need to be strong. Debonding at the interface, internal cavitation of the toughener phase, and microcracking and crazing of the inclusion phase can effectively serve the purpose of shielding the crack and impeding crack growth. The preceding dilatational processes can also result in matrix microcracking, depending on the physical nature of the matrix material. Consequently, a toughened system is obtained.

For the crack deflection and crack bifurcation mechanisms to occur, the toughener phase needs to generate sufficient stress disturbance in front of the crack tip. When the crack propagates, the crack path will then be altered by the stress disturbance. This causes crack deflection and crack bifurcation mechanisms to occur. The rubber particle size also appears to play an important role in deflecting the crack. Comparing Fig. 6 with Fig. 8 and 10 clearly shows that a larger rubber particle (or a hole) can deflect the crack more effectively than a smaller one. Further, when the smaller particles cluster together, they can be mechanically treated as a big particle (i.e., the CSR-C, CSR-F, and CSR-G systems). Consequently, the crack deflection mechanism can be enhanced. Therefore, a larger rubber particle size or a local clustering of small particles maintaining a good global particle dispersion should be utilized when the crack deflection mechanism is to be promoted.

For the crack pinning mechanism to take place, the toughener phase needs to adhere to the matrix strongly. Thus, when the crack grows around the particles, the crack front is bowed and more energy is required to propagate the crack. It has been shown that the particle size and concentration are critical in toughness optimization via crack pinning [56]. The maximum stress concentration around a rigid inclusion is at the polar region, thus, when the crack grows slowly, it will tend to grow toward the polar region of the particle. This discourages occurrence of the crack pinning mechanism. Only when the crack grown quickly, which is usually true for a brittle epoxy matrix, will it grow into the rigid toughener particles and possibly cause crack pinning to occur. When the toughener particle is soft, as in CTBN rubber particles, and when the matrix does not undergo shear yielding, the crack bridging mechanism is also likely to take place right after the pinning mechanism occurs. When the toughener particles are softer than the matrix and are not stretchable, like the Proteus particles [23], then the crack bridging mechanism becomes inoperative and the crack pinning mechanism becomes dominant.

When failure occurs in epoxy systems, usually more than one toughening mechanism is taking place. In order to optimize the toughening effect, it is desirable to promote more than one toughening mechanism when fracture occurs. Therefore, it is important to first consider promoting the most effective toughening mechanisms, such as shear yielding and croiding, and then, without sacrificing the above, promote other effective mechanisms. This concept can best be demonstrated by the CSR-G–modified epoxy system where the rubber cavitation and extended matrix shear yielding mechanisms are preserved, while an additional crack deflection mechanism is promoted through the local clustering of the CSR-G particles.

Furthermore, since practically all the operative toughening mechanisms in rubber-modified epoxies are viscoelastic and viscoplastic, they are time and temperature dependent. The toughening approach, for a given material, may as a result vary from one testing condition to another. Therefore, caution should be exercised in any attempt to apply fundamental knowledge learned in toughening under one testing condition to another.

Finally, the mechanics tools introduced in this chapter are quite useful in gaining fundamental understanding of how and why certain toughening mechanisms are affected by the morphology of the toughened system. The importance of relieving the plane strain constraint in triggering matrix shear yielding can also be rationalized via the mechanics approach. Detailed descriptions on how to utilize these mechanics tools effectively is not covered in this chapter because of the complexity and the stand-alone nature of this subject. They can be found elsewhere [27,34–40].

VI. CONCLUDING REMARKS

This chapter, while concentrating on the recent experimental efforts conducted by the authors in toughening brittle epoxies, focuses on using the more generic materials science and mechanics tools to study routes for toughening epoxies. This will inevitably help provide a broader utilization of the present understanding in toughening other types of polymers as well as utilization in other applications. Effects due to rubber type, particle size, interfacial adhesion, and particle dispersion in epoxy toughening are discussed. Methodology regarding how to make a tough epoxy system is also addressed. It should be cautioned that in the field of polymer toughening, many toughening concepts and theories are still tentative. Only the experimental data are reliable. Care needs to be taken when utilizing any toughening theories and moldes to design new products.

ACKNOWLEDGMENTS

The authors would like to thank Professor A. F. Yee for his constant valuable discussions concerning this work. The authors also would like to thank Professor R. S. Porter and Ms. C. Stamm for permitting the use of work previously published in *Polymer Engineering and Science,* and the American Chemical Society Book Series No. 233 (*Toughened Plastics: Science and Engineering*). Special thanks are given to R. E. Jones, D. L. Barron, C. C. Garrison, N. A. Orchard, D. W. Hoffman, C. E. Allen, C. Bott, R. D. Peffley, L. M. Kroposki, and T. E. Fisk for their input, experimental assistance, and material supplied for this work.

REFERENCES

1. A. J. Kinloch, C. A. Finch, and S. Hashemi, *Polym. Commun.* 28:332 (1987).
2. M. D. Glad, Ph.D. thesis, Cornell University, Ithaca, NY, 1986.
3. J. N. Sultan and F. J. McGarry, *J. Polym. Sci.* 13:29 (1973).
4. W. D. Bascom, R. L. Cottingham, R. L. Jones, and P. Peyser, *J. Appl. Polym. Sci.* 19:2545 (1975).
5. S. Kunz-Douglass, P. W. R. Beaumount, and M. F. Ashby, *J. Mater. Sci.* 16:2657 (1981).
6. S. Kuntz, Ph.D. thesis, University of Cambridge, Cambridge, England, 1978.
7. R. A. Pearson and A. F. Yee, *J. Mater. Sci.* 26:3828 (1991).
8. A. F. Yee and R. A. Pearson, *J. Mater. Sci.* 21:2462 (1986).
9. R. A. Pearson and A. F. Yee, *J. Mater. Sci.* 24:2571 (1989).
10. A. J. Kinloch, S. J. Shaw, D. A. Tod and D. L. Hunston, *Polymer* 24:1341 (1983).
11. I. M. Low and Y. W. Mai, *Compos. Sci. Technol* 33:191 (1988).
12. G. Levita, in *Rubber-Toughened Plastics*, (C. K. Riew, ed.), Advances in Chemistry Series No. 222, American Chemical Society, Washington, DC, 1989, p. 93.
13. A. J. Kinloch, in *Toughened Plastics*, (C. K. Riew, ed.), Advances in Chemistry Series No. 222, American Chemical Society, Washington, DC, 1989, p. 67.
14. R. A. Pearson and A. F. Yee, *J. Mater. Sci.* 21:2475 (1986).
15. A. F. Yee, R. A. Pearson and H.-J. Sue. *7th Int. Conference on Fracture 4:*2739 (1989).
16. D. S. Parker, H.-J. Sue, J. Huang, and A. F. Yee, *Polymer* 31:2267 (1990).
17. H.-J. Sue and A. F. Yee, *J. Mater. Sci.* 24: 1447 (1989).
18. D. E. Henton, D. M. Pickelman, C. B. Arends, and V. E. Meyer, U.S. Patent 4,778,851, 1988.
19. D. K. Hoffman and C. B. Arends, U.S. Patent 4,708,996, 1987.
20. D. K. Hoffman and C. B. Arends, U.S. Patent 4,789,712, 1988.
21. H.-J. Sue, *Polym, Eng. Sci.* 31:270 (1991).
22. H.-J. Sue, R. A. Pearson, D. S. Parker, J. Huang, and A. F. Yee, *ACS Polym. Preprint 29:*147 (1988).
23. H.-J. Sue, *Polym. Eng. Sci.* 31:275 (1991).

24. H.-J. Sue, E. I. Garcia-Meitin, B. L. Burton, and C. C. Garrison, *J. Polym. Sci. Polym. Phys. 29:*1623 (1991).

25. A. C. Roulin-Moloney, *Fractography and Fracture Mechanisms of Polymers and Composites.* Elsevier Applied Science Publishers, New York, 1989.

26. L. C. Sawyer and D. T. Grubb, *Polymer Microscopy,* Chapman & Hall, New York, 1987.

27. D. Broek, *Elementary Engineering Fracture Mechanics,* Noordhoff International, The Netherlands, 1974.

28. J. F. Knott, *Fundamentals of Fracture Mechanics,* Butterworth, London, 1976.

29. Y. Mura, *Micromechanics of Defects in Solids,* Martinus Nijhoff, Boston, 1982.

30. A. G. Evans, Z. B. Ahmad, D. G. Gilbert, and P. W. R. Beaumont, *Acta Metall, 34:*79 (1986).

31. O. C. Zienkiewicz, *The Finite Element Method,* McGraw-Hill, New York, 1977.

32. H.-J. Sue and A. F. Yee, *Polymer 29:*1619 (1988).

33. H.-J. Sue, R. A. Pearson, and A. F. Yee, *Polym. Eng. Sci. 31:*793 (1988).

34. H.-J. Sue, Ph.D. thesis, The University of Michigan, Ann Arbor, 1988.

35. W. Johnson and P. B. Mellor, *Engineering Plasticity,* Van Nostrand Reinhold, New York, 1973.

36. R. Hill, *Plasticity,* Clarendon Press, Oxford, 1950.

37. G. R. Irwin. Trans, Am. Soc. Mech. Eng., *J. Appl. Mech. 24:*361 (1957).

38. J. N. Goodier, *J. Appl. Mech. 1:*39 (1933).

39. J. J. Dewey, *Appl. Phys. 18:*578 (1947).

40. J. D. Eshelby, *Proc. R. Soc. (Lond.) A241:*376 (1957).

41. P. C. Yang, E. P. Woo, H.-J. Sue, M. T. Bishop, and D. M. Pickelman, *ACS PMSE, 63:*315 (1990).

42. A. F. Yee, Modifying Matrix Materials for Tougher Composites, *Toughened Composites ASTM STP 937* (N. Johnston, ed.), American Society for Testing and Materials, Philadelphia, 1986, p. 377.

43. G. B. Portelli, W. J. Schultz, R. C. Jordan, and S. C. Hackett, *Polym. Comp. 2:*381 (1989).

44. S. Tong, C. Chen, and P. T. K. Wu., in *Rubber-Toughened Plastics,* (C. K. Riew, ed.) Advances in Chemistry Series No. 222, American Chemical Society, Washington, D.C., 1989, p. 376.

45. D. Verchere, J. P. Pascault, H. Sautereau, and S. M. Moschiar, *J. Appl. Polym. Sci. 43:*293 (1991).

46. H.-J. Sue, J. Huang, and A. F. Yee, *Polym. Commun. 83:*4868 (1992).

47. S. Wu, *Polymer 26:*1855 (1985).

48. A. C. Garg and Y. W. Mai, *Compos. Sci. Technol. 31:*179 (1988).

49. H.-J. Sue, *J. Mater. Sci. 36:*1395 (1991).

50. H.-J. Sue, E. I. Garcia-Meitin, and N. A. Orchard. *J. Polym. Sci. Polym. Phys. Ed. 31:*595 (1993).

51. A. G. Evans and K. T. Faber, *J. Am. Ceram. Soc. 67:*255 (1984).

52. K. T. Faber and A. G. Evans, *Acta Met. 31:*565 (1983).

53. K. T. Faber and A. G. Evans, *Acta Met. 31:*577 (1983).

54. A. B. J. Clark and G. R. Irwin, *Exp. Mech. 6:*321 (1966).

55. M. Ramulu and A. S. Kobayashi, *Exp. Mech. 23:*1 (1983).

56. F. F. Lange and K. C. Radford, *J. Mater. Sci. 6:*1199 (1971).

57. F. F. Lange, Fracture of Brittle Matrix Particulate Composites, *Composite Materials, Vol. 5: Fracture and Fatigue* L. J. Broutman, ed.), Academic Press, New York, 1974, p. 2.

6
Rubber-Toughened Thermosets

FREDERICK J. McGARRY Massachusetts
Institute of Technology, Cambridge,
Massachusetts

I. INTRODUCTION

The desire to toughen thermoset polymeric glasses, polyesters, and epoxies arose from a practical need: when reinforced with a woven fiberglass fabric (style 181 weave) and tested under tension, such laminates displayed a break or knee in the stress–strain curve, a point where the modulus decreased significantly. This was shown to be the stress at which internal cracking of the resin matrix started and proliferated; below the threshold stress, the laminate maintained its integrity, in both static and cyclic loading, whereas above it, progressive cracking deterioration took place [1]. Since the knee was found at only 20 to 30% of the ultimate tensile strength, the safe working stress level to avoid internal cracking was a small fraction of the laminate's ultimate capacity. If the knee could be raised, if the matrix could be made more resistant to crack

175

initiation, a larger fraction of the laminate strength could be utilized. A tougher matrix would provide a structurally more efficient laminate.

Earlier research with cast polymethyl methacrylate (PMMA) had shown that the fracture surface work term γ in the Griffith equation was several orders of magnitude larger than that associated with simple chain scission: 10^5 to 10^6 ergs/cm^2 compared to the chain scission model of approximately 10^3 ergs/cm^2 [2]. The discrepancy was caused by craze formation and deformation, mechanisms intensively studied by Kambour [3] and later by Kramer [4]. Even when the PMMA was chemically crosslinked enough to inhibit or suppress obvious craze formation, the γ value dropped only by a factor of 10, to the range of 10^4 to 10^5 ergs/cm^2, a value that was still appreciably greater than the chain scission model value [2]. In the crosslinked system, substantial plastic flow still took place on a molecular level at the crack tip. Other, subsequent experiments with crosslinked epoxies and polyesters showed a similar behavior [5], and the insightful observations by Bucknall and Smith (6) on rubber-toughened polystyrene, which were published about that time, suggested that rubber particles might also toughen a crosslinked epoxy or polyester. (It was not until much later that Glad and Kramer conclusively showed that the crosslinked epoxy shear yields rather than crazes [7], so the rubber–epoxy, and rubber–polyester work went forward on an incorrect premise, though to a successful conclusion.)

A number of preformed rubbery particles were tried, none with much success, but then J. Carey, of Shell Chemical, brought to my attention the spectrum of (reactive) liquid rubbers manufactured by BF Goodrich for use as binders for solid rocket propellants. These low molecular weight, butadiene-acrylonitrile copolymers were soluble in liquid diglycidyl ether of bisphenol A (DGEBA) epoxy and could be formulated with amine and carboxyl groups both at the chain ends and pendant along the chain. When the epoxy–rubber solution was cured using conventional procedures, rubber particles precipitated out as a second phase, with very little reduction in the glass transition temperature, the modulus, or the tensile strength of the epoxy. The fracture toughness increased, somewhat dramatically, however: γ values of 10^5 to 10^6 ergs/cm^2 were obtained with as little as 10 phr rubber in 100 parts of epoxy.

Work with DGEBA epoxy (Shell Chemical Epon 828) and CTBN rubber (carboxy-terminated butadiene-acrylonitrile) was pursued vigorously. The effects of such parameters as cure temperature, initial molecular weight of the rubber, butadiene-acrylonitrile ratio, terminal and pendant reactive groups, catalyst type, and concentration were studied [8]. A number of important principles became clear:

1. Fracture toughness was proportional to rubber content, up to 25 phr rubber.
2. Particles in the range of 1 to 2000 nm gave the most effective toughening.
3. The particles were not pure rubber; substantial amounts of epoxy were present, both chain extending and crosslinking the rubber.
4. The particles produced unusually large cavities in the fractured surfaces of the epoxy, under slow crack propagation. The spherical cavities were several times larger than the diameter of the particles producing them.
5. Such cavities retracted to a planar morphology when the fracture surface was raised above its glass transition temperature. Substantial plastic flow had taken place in the epoxy when the material cavitated.
6. Chemical bonding between the particles and the matrix was necessary for toughening; when it was absent, no effect was achieved.
7. A bi-distribution of particle sizes provided the most efficient use of the rubber, if the size peaks were located at 200 and 2000 nm the toughness per unit of added rubber was greater than for a monodispersion of either.

8. The increased toughness was rate and temperature sensitive and was lost at high crack speeds or low testing temperatures.
9. Toughened matrices reinforced with the 181 fiberglass fabric could be taken to much higher tensile stress levels before internal cracking occurred, in both static and cyclic loading.

II. SUBSEQUENT DEVELOPMENTS

After the work described above was published, many investigators started exploring the area in detail. Probably the most convenient record of their works is the series of American Chemical Society (ACS) monographs edited by Riew and colleagues, which continues to the present time [9,10,11]. In this narrative I shall mention only a few of these works, which I believe to be the most significant derivatives of the original finding.

How could a presumably immobile, crosslinked glassy thermoset be toughened by rubber particles? How could the birefringence effects at the crack tip plastic zone be so similar to those reported by Bucknall and Smith [6] for high-impact polystyrene (HIPS)? The first serious attempt to explain this was made by Kunz-Douglass and colleagues, who postulated that the rubber particles deformed and extended while remaining bonded to the matrix, and thereby absorbed substantial mechanical work before the crack could advance [12,13]. Rubber ligaments would bridge the crack until they broke. Although this was physically plausible, and in fact could be observed, the model was not quantitatively consistent: the rubber deformation to fracture work simply did not represent a large enough energy sink.

Concurrently, Bascom and his colleagues pursued the technology of rubber-toughened epoxies in metal-to-metal adhesive joints, of great interest to the aircraft and automotive industries [14,15] In fact, it is in this application that the idea has had its greatest value. Most engineering joints contain large peel stresses, and their breaking loads are directly dependent upon the resistance of the adhesive to opening-mode crack initiation and propagation. Breaking strength is proportional to adhesive fracture toughness.

Bascom and Hunston used an aluminum, tapered double cantilever beam specimen held together with DGEBA epoxy adhesive in which the CTBN content varied from 0 to 18.5 phr; cast bulk samples of the same compositions were also made and tested. Increases in the fracture energy of the bulk samples were paralleled by increases in the adhesive; effectively a 1:1 correlation was found. When the rubber content was held at 18.5 phr and the bond thickness was varied from nearly 0 to about 3 mm, the fracture energy rose rapidly to a peak at around 0.5 mm (4 kJ/m^2), then as rapidly declined to about half the peak value at 1.0 mm, remaining constant thereafter. (The same adhesive without rubber had a fracture energy of about 0.1 kJ/m^2 which did not vary with bond thickness.) Later Hunston [16] and Wang [17] showed that the peak was a kind of artifact: as the bond thickness decreased, the tensile stresses at the crack tip increased, thereby increasing the plastic zone size and the fracture energy. Eventually, as the thickness further decreased, the metallic adherends inhibited the plastic zone and the fracture energy decreased.

Rate and temperature studies of fracture energy on bulk samples produced data that could be rationalized by a time–temperature superposition treatment. This suggested that the fracture energy absorption was viscoelastic in origin, being based upon flow phenomena taking place in the system. The significance of such a finding may not have been widely appreciated at the time.

One of the most persistent and productive researchers of rubber-toughened epoxies has been Kinloch, who did follow the lead provided by Hunston's work. Kinloch showed that the rubber particles initiated shear bands or deformation zones by causing a stress concentration in the surrounding glassy matrix. These deformations tend to localize because of strain softening in the epoxy and because they encounter other particles than the initiating one; usually a band will start at one particle and terminate at another one [18]. Thus if many particles are present, the toughness improves because more deformation zones are created and more are terminated before causing fracture. A larger fraction of the glassy phase undergoes shear yielding. Kinloch also directed a great deal of his attention to the cavitation effect in the matrix around a particle and related the size of the cavities to the macroscopic ductility of the matrix material, the latter being subject to manipulation by crosslink density and/or testing temperature.

During this same period of time three other researchers contributed greatly to our knowledge of rubber-toughened thermoset polymer glasses: Siebert, Riew, and Rowe. Because of their employment at the producer of reactive liquid rubbers (BF Goodrich) they were in an unusually favorable position to research the area, and many advances were achieved. Principal among these were the synthesis of rubbers with amine, epoxy, and various vinyl functionalities [19], the development of prereacted formulations that provided better control over particle size and composition [20], the refinement of toughening systems for use in unsaturated polyester systems [21], extensive studies of the effects of various curative agents [21], explorations of core–shell particles [22], and valuable research on the effects of bonding at the particle–matrix interface [23]. In addition, over the years they have provided generous supplies of reactive liquid polymers to other researchers all over the world; often these had special characteristics that helped to advance the state of the art.

As mentioned earlier, adhesives technology was the first principal target of rubber toughening that benefited immediately and substantially. Another, later one, has been high-performance aircraft laminates, usually composed of continuous carbon fibers in highly crosslinked epoxy resins. The high crosslink density is necessary to retain good properties under "hot, wet" conditions: higher temperatures (200 to 300 °F) and 100% RH. Coupled with this requirement is the need to show minimum internal delamination damage when the plate is struck by a foreign object; the delamination lowers the in-plane compressive strength, reducing the plate's structural capacity. Evans and his colleagues were the first to show that the damage depended upon the mode II fracture toughness of the matrix and that this, too, could be increased by including appropriate rubber particles in the epoxy [24]. With the rubber present, however, the hot, wet properties of the composite declined to an unacceptable degree, so to avoid this loss they conceived the idea of a very thin, tough resin film interleaved between each pair of graphite–epoxy plies [25]. The film worked very well; it produced high resistance to impact damage while retaining the good hot, wet performance of the highly crosslinked epoxy matrix. Initially the film was rubber-epoxy; later versions used high-temperature thermoplastic films, and current practice sprinkles high-temperature thermoplastic particles on one side of each layer of prepreg before the assembly is cured under heat and pressure. This facilitates the retention of a high-fiber-volume fraction in the composite, a factor of great importance in aircraft applications.

As the demand for higher-temperature performance epoxies has increased, the effectiveness of reactive liquid rubber additions has diminished for two reasons: (1) the low T_g rubber lowers the maximum use temperature and the modulus of the epoxy and (2) the high crosslink density in the epoxy reduces its ability to shear-deform and thus the toughening effect of the rubber particles is lessened. With respect to the latter point, Fischer in an extraordinarily comprehensive review, showed how crosslink density decreases the crack tip plastic zone size and the crack opening displacement (COD): the opening-mode critical stress intensity factor is a strong function of the molecular weight between crosslinks [26]. With a DGEBA epoxy having a T_g of 212°C, the K_{1C} value is 0.75 MPa $m^{1/2}$ and the COD is 0.94 μm. Rubber particle toughening of such a network is, at best, marginally effective. The alternative now being explored vigorously has been mentioned: high-temperature thermoplastics such as polyether sulfone (PES), modified PES, polyether inside (PEI), polyether ether ketone (PEEK), and many other polyimides. These are used in epoxies in two ways: (1) as finely granulated particles added either to the liquid epoxy or to the B-staged material; in both cases the particles retain their original identity or (2) as a polymer dissolved in the liquid epoxy and later precipitated out as a second-phase particles; often the solubility is achieved by modifying the polymer with epoxy groups. The modulus and T_g of the epoxy are maintained and the toughness is improved in direct proportion to the amount of thermoplastic that is added. For example, Hedrick et al reported a doubling of K_{1C} (from 0.6 to 1.2 $Nm^{3/2}$) in an epoxy having a T_g of 195°C when 15% of PSF was present in approximately 1.0 μm-diameter particles [27]. They found that the PSF had to be above a certain molecular weight to achieve such toughening, which strongly suggests the mechanism is principally crack pinning: the tough particles impede the advance of the crack and their ductility causes them to absorb energy before fracturing and permitting the crack to grow. It would be expected that interfacial adhesion would be important to the process and on this point their data are contradictory; one PSF system they investigated produced toughening without apparent adhesion, whereas a second, also effective, showed good interfacial adhesion. The morphologies of the two did differ, however.

There is another feature of the epoxy–thermoplastic systems that is extremely interesting and that has been pursued. At about 20 to 25% thermoplastic content, a co-continuous morphology is formed, with both the epoxy and the thermoplastic in a kind of interpenetrating network structure that exhibits maximum fracture toughness. (At higher contents, because of a phase inversion the thermoplastic becomes the

continuous phase, filled with epoxy domains or particles.)
Almen et al. have used modified PES in epoxy to produce
carbon fiber laminates with such matrices, and they have
been found to have superior hot–wet and compression-after-
impact properties [28]. In fact, many other people have
worked with epoxy–thermoplastic blends; commercial products based on them are available and they are used in some
applications, but processing constraints and cost factors have
influenced their acceptance [29].

III. CURRENT UNDERSTANDING

Returning to the epoxy–rubber systems, which still offer the
highest fracture energy values of any available, we have the
question of where the toughness comes from: (1) rubber ligaments bridging the crack? (2) shear plastic flow in the epoxy?
(3) or the growth of the cavities in the epoxy initially occupied
by the rubber particles? Kinloch first modeled the system in
these three terms (30) and subsequently Huang, Kinloch, and
colleagues (31) made the analysis of the model more comprehensive, using it both for predictive purposes and to identify
the relative contributions of the three mechanisms. For example, at room temperature conditions their model predicted a
G_{1C} value of 4.8 kJ/m^2 versus 5.9 measured, with 8% of the
latter due to rubber bridging, 54% due to shear banding, and
38% represented by cavity growth. Except at very low temperatures, less than $-20°C$, where cavitation ceases, these
relative contributions are representative; shear yielding and
cavity growth are the dominant energy-absorbing mechanisms, and the latter can occur because the rubber particles
pull apart under the two actions of residual thermal stresses
and the externally applied load (increased by the local stress
concentration due to the elastic discontinuity). The model is
persuasive and until a better one is advanced, it should be
viewed as the most successful to date.

Another matter, mentioned earlier, is worth reconsidering. Long ago it was found that a bidistribution of rubber particle sizes results in a disproportionally great increase in the fracture toughness of an epoxy matrix. The crack tip plastic zone increases in size and the toughness does also. (The same is true of ABS and HIPS systems.) This has been done by numerous investigators, and several different routes to produce the bidistribution have been used; it appears to be not route specific. Recently we worked with non-phase-separated alloys of DGEBA epoxy and amine-terminated butadiene acrylonitrile (ATBN) and found that even small rubber contents coreacted with the epoxy significantly reduces its yield strength and increases its tensile extensibility; it effectively plasticizes the glassy epoxy (32). In a particle-toughened system, it is very unlikely that the boundary region around each particle has a compositional discontinuity; much more probable is a gradient from the rubber-rich to the glass-rich over a distance of a few hundred Angstroms. In this region the shear yield stress of the epoxy is lowered and shear band formation is enhanced, which probably also is the case with the "normal," monosized particles. Thus, if a second population of much smaller particles is also present, the substantially greater volume of locally plasticized epoxy is activated by the bands forming at the large particles and a more global yielding takes place; the toughness rises remarkably. Were only the small particles present, the shear band generation would be much less, and the toughness lower. Thus, I speculate that the main role of the little particles is to increase the volume of interphasal material, a very locally plasticized epoxy glass.

IV.   NEW DEVELOPMENTS

Because of cost, much more polyester than epoxy resin is used in composites. Since the crosslink density is high in the styrene monomer–unsaturated polyester systems, rubber particle toughening of them has been much less effective: 10 parts of rubber in epoxy can increase its G_c by a factor of 20 to 40, whereas a similar concentration (usually of a different composition rubber) in polyester will produce only a fivefold increase. Recently, we have found what may be a way around this constraint, without compromising the modulus and HDT of the polyester too badly. Four principal ingredients are involved: unsaturated polyester, monomeric styrene, DGEBA epoxy, and ATBN. With appropriate peroxide and amine catalysts, castings cured at moderate temperatures and times (150°C 30 min) result in clear plates with enhanced toughness proportional to the ATBN content. The ratio of polyester to epoxy can range from 4:1 to 1:1, and rubber contents as high as 25% have been explored. At concentrations of less than about 10% ATBN, the cured structure is composed of a discontinuous rubber-modified glass in a continuous glassy phase. However, the volume fraction of the discrete phase is 50 to 60%. Above 10% rubber, a phase inversion takes place; now the continuous phase is rubber modified and the discontinuous one is clear glass. In both regimes, transmission electron microscopy of stained sections shows the rubber domains to be only a few thousand Angstroms in size. Glass-fiber-reinforced composites based on this system have been made and tested in Gardner impact, to observe the damage area and volume as functions of rubber content. Both decrease as the rubber content increases (33).

V. CONCLUDING COMMENTS

The technology of rubber-toughened epoxies, nearly 30 years old, has received much attention and development from a number of people. Commercial adhesives, sealants, and laminating resin systems using the idea now are widely available. It has been extrapolated to use tough, temperature-resistant thermoplastics, both as separate phase particles and as co-continuous networks in more highly crosslinked epoxies. The actions of the rubber particles in epoxy are quite well understood; the behavior of the thermoplastic-modified systems is still being studied and debated. Just recently a novel system combining reactive rubber, epoxy, and polyester has been identified and it appears to have considerable potential for improving the damage resistance of fiber-reinforced polyester articles. It will be interesting to see how it develops in the years ahead.

REFERENCES

1. F. J. McGarry, Resin Cracking in Composites, Chem. Eng. (Institute of Chemcial Engineers, London) 42 (8): (1964).
2. L. J. Broutman, F. J. McGarry, Fracture Surface Work Measurements on Glassy Polymer by a Cleavage Technique, I and II, *J. App. Poly. Sci. 9:589, 609* (1965).
3. R. P. Kambour, *A Review of Crazing and Fracture in Thermoplastics.* Report 72CRD285, General Electric Corporation R&D, Schenectady, NY, 1972.
4. E. J. Kramer, Mechanisms of Toughening in Polymer Mixtures, *Mechanisms of Toughening in Polymer Mixtures, Polymer Compatability and Incompatability: Principles and Practice* (K. Solc, ed.), Horwood Academic Publishers, New York, 1983.
5. F. J. McGarry, Crack Propagation Resistance of Thermosetting Resins, Proc. 148 Nat. Mtg. ACS. Div. Org. Ctgs. and Plas. Chem., Vol. 24, No. 2, Chicago, 1964.
6. C. B. Bucknall and R. R. Smith, *Polymer 6:*437 (1965).
7. M. D. Glad, Microdeformation and Network Structure in Epoxies, Ph.D. thesis, Cornell University, Ithaca, NY, 1986.
8. F. J. McGarry, A. M. Willner, and J. N. Sultan, *Toughening of Glassy Crosslinked Polymer with Elastomeric Inclusions,* Civ. Eng. Dept., Report R69-35, Massachusetts Institute of Technology, Cambridge, 1969.
9. C. K. Riew and J. K. Gillham, Eds. *Rubber Modified Thermoset Resins,* Advances in Chemistry Series No. 208, American Chemical Society, Washington, DC, 1984.

10. C. K. Riew, ed., *Rubber Toughened Plastics*, Advances in Chemistry Series No. 222, American Chemical Society, Washington, DC, 1989.

11. C. K. Riew and A. J. Kinlock, eds., *Toughened Plastics I: Science and Engineering*, Advances in Chemistry Series No. 233, American Chemical Society, Washington, DC, 1993.

12. S. Kunz-Douglass, P. W. R. Beaumont, and M. F. Ashby, *J. Mater. Sci. 15:*1109 (1980).

13. J. A. Sayer, S. C. Kunz, and R. A. Assink, *Polymer Preprints, ACS Div. Polym. Mater. Sci. Eng. 49:*442 (1983).

14. W. D. Bascom and D. L. Hunston, in *Rubber Modified Thermoset Resins* (C. K. Riew and J. K. Gillham, eds.) Advances in Chemistry Series No. 208, American Chemical Society, Washington, DC, 1984, p. 83.

15. W. D. Bascom and D. L. Hunston, *Rubber Toughened Plastics* (C. K. Riew, ed.), Advances in Chemistry Series No. 222, American Chemical Society, Washington, DC, 1989, p. 135.

16. D. L. Hunston, A. J. Kinloch, S. J. Shaw, and S. S. Wang, *Adhesive Joints*, (K. L. Mittal, ed.), Plenum Press, New York, 1984, p. 789.

17. S. S. Wang, J. F. Mandell, and F. J. McGarry, *Int. J. Fracture 14:*39 (1978).

18. A. J. Kinloch, S. J. Shaw, and D. L. Hunston, *Polymer 24:*1355 (1983).

19. A. R. Siebert, in *Rubber Modified Thermoset Resins*, (C. K. Riew and J. K. Gillham, eds.) Advances in Chemistry Series No. 208, American Chemical Society, Washington, DC, 1984, p. 179.

20. R. S. Drake and A. R. Siebert, *SAMPE Quart. 6(4):*7 (1975).

21. A. R. Siebert, C. D. Guiley and A. M. Eplin, in *Rubber Toughened Plastics* (C. K. Riew, ed.), Advances in Chemistry Series No. 222, American Chemical Society, Washington, DC, 1989, p. 389.

22. C. K. Riew and R. W. Smith, in *Rubber Toughened Plastics* (C. K. Riew, ed.), Advances in Chemistry Series No. 222, American Chemical Society, Washington, DC, 1989, p. 225.

23. A. R. Siebert and C. K. Riew, *Org. Coat. Plast. Chem. 31:*552 (1971).

24. R. Krieger, Proc. 29th Natl. SAMPE Symposium, 1984.

25. R. E. Evans, J. E. Masters, and J. L. Courter, Proc. Adv. Comp. Conf., ASM, Dearborn, MI, 1985.

26. M. Fischer, *Adv. Polym. Sci. 100:*313 (1992).

27. J. L. Hedrick, I. Yilgor, M. Jurek, J. C. Hedrick, G. L. Wilkes, and J. E. McGrath, *Polymer. 32:*2020 (1991).

28. G. R. Almen, R. M. Byrens, P. D. McKenzie, R. K. Maskell, P. T. McGrail and M. S. Sefton, Proc. 34th Intl. Symp. SAMPE, Vol. 34, Book 1, 1989, p. 259.

29. H. G. Recker, T. Allspach, V. Altstaedt, T. Folda, W. Heckman, P. Ittelmann, H. Tesch, and T. Weber, Proc. 34th Intl. Symp. SAMPE, Vol. 34, Book 1, 1989, p. 747, A. R. Wedgewood and D. C. Grant, Proc. 34th Intl. Symp. SAMPE, 1989, Vol. 34, Book 1, p. 1.

30. A. J. Kinloch and R. J. Young, *Fracture Behavior of Polymers*, Applied Science Publishers, London, 1983, p. 303.

31. D. D. Huang et al., in *Rubber Toughened Plastics* (C. K. Riew, ed.), Advances in Chemistry Series No. 222, American Chemical Society, Washington, DC, 1989, p. 119.

32. F. J. McGarry, in *Rubber Toughened Plastics* (C. K. Riew, ed.), Advances in Chemistry Series No. 222, American Chemical Society, Washington, DC, 1989, p. 173.

33. R. Subramaniam, F. J. McGarry, and P. H. Flueler, Damage Resistant SMC, Proc. 2nd Intl. Conf. on Deformation and Fracture of Composites, Mat. & Sci, Div., Inst. of Materials, Manchester, UK, 1993, p. 11–1.

7
Polyethylene

GEORGE W. KNIGHT   The Dow Chemical
Company, Freeport, Texas

I. INTRODUCTION

"It is the simplest of polymers, it is the most complex of polymers." This paraphrase of Dickens introduction to *A Tale of
Two Cities* aptly describes polyethylene. It is chemically very
simple, being comprised of only carbon and hydrogen (ignor-
ing nonhydrocarbon copolymers such as ethylene vinyl acetate, ethylene acrylic acid, etc.) and yet, because the carbon
and hydrogen can be arranged in an almost infinite number of
ways, its physical structure is extremely complex. Because of
this, there is no single description of polyethylene.

Without copolymerization of other α-olefins with ethylene
two very different types of ethylene polymers can be made:
low-density polyethylene, which is made at very high pressures using a free-radical initiator, and high-density polyethylene, which is made using various metal-complex catalysts.

---

*Retired

The low-density polyethylene (LDPE) contains many short branches, mostly two to four carbons long and a few long branches of undetermined length, but long enough to have a significant effect upon the melt properties, viscosity and elasticity. High-density polyethylene (HDPE) theoretically has no branches (although a few may be present due to oligomers formed by the catalyst and then copolymerized with the ethylene). Copolymerization of α-olefins with ethylene, usually from C_3 to C_8 in length, gives a whole new family of polyethylenes commonly referred to as linear low-density polyethylene (LLDPE).

Polyethylene is a semicrystalline solid and the degree of crystallinity is dependent upon defects on the polymer chain that interfere with the ability of the chain to be arranged in a crystal. These defects are mainly branch points on the backbone of the polymer chain, which are formed by both inter- and intramolecular mechanisms in the LDPE radical polymerization process, and by copolymerizing with hydrocarbon α-olefins in the HDPE transition metal–catalyzed polymerization processes. The percent crystallinity, or density, has a major effect upon the physical properties of PE, and controlling the length and quantity of the branches controls the percent crystallinity of the polymer. Because of the high degree of complexity of polyethylene and the versatility in controlling this complexity, polyethylene can be designed to take advantage of a large variety of properties that are inherent within its molecular structures. All properties of polyethylene are determined by the molecular weight (MW), molecular weight distribution (MWD), branching (both short and long) and branching distribution (both inter and intra). Nearly all attempts to toughen polyethylene are made by changing one of these parameters. Other means often used to make "tougher" polyethylene include modifying the basic composition of the polymer by copolymerizing with nonhydrocarbon monomers (e.g., vinyl acetate, acrylic acid, methacrylic acid, etc), modifying the polyethylene via crosslinking with peroxide, silanes, or irradiation (gamma or beta), or blending various types of fillers into the polyethylene.

II. APPLICATIONS OF POLYETHYLENE

Unlike many polymers the applications of polyethylene can be divided into two major areas, durables and nondurables. Durables are products such as wire and cable coatings, agricultural tanks, some large containers, or others products that must withstand a relatively long service life. On the other hand, nondurables are usually for a one-time, short-term application, such as packaging films, milk bottles, paper coating, and so on.

The applications of polyethylene dictate the reasons for wanting to improve its strength. Durables need to be tougher in order to endure for a long period of time. For example, wire and cable resins are expected to last a minimum of 20 years in outside service, often at elevated temperatures. Agricultural and other storage tanks must maintain modulus, impact strength, and so on, in severe environments. Therefore, durables are required to maintain strength, environmental stress crack resistance (ESCR), creep resistance, and overall toughness throughout their intended lifetime. On the other hand, toughness in polyethylene used in nondurable applications is desirable mainly to allow down-gauging, which reduces the amount of material required to make the product. This results in savings for the converter and therefore gives an advantage to the polyethylene producer who is able to produce the toughest polymer. The introduction of linear low-density polyethylenes (LLDPE) led to significant advantages in both durables and nondurables, because they give long-term strength and also allow a high degree of film down-gauging and thinner molded parts.

III. STRUCTURES OF POLYETHYLENE

Polyethylene is not commonly toughened by the addition of other substances, but is modified to take advantage of its inherent "self-toughening" structure; that is, it is a mixture of crystalline and amorphous regions that can be modified and adjusted to take advantage of the best properties of each. Since the strength of polyethylene is mainly a function of its complex structure, and since the main method of improving the strength of polyethylene is by modifying its structure, some time will be devoted to a discussion of its structure and the effects of modifying different structural aspects of the various types of polyethylenes.

Consideration of the crystalline domains has long dominated the research to characterize and explain the physical properties of semicrystalline polymers such as polyethylene. It has been relatively easy to study the crystalline regions, both qualitatively and quantitatively, using wide-angle and small angle x-ray, infrared spectroscopy, low-angle laser light scattering, and electron microscopy. However, the noncrystalline portion of polyethylene is equally important, or even more so, in determining its strength properties. The noncrystalline portion is really two regions, the amorphous and the interfacial. The interfacial region is the transition region between the crystalline and amorphous regions and has some characteristics of each. Much research has been done in the last few years to better define and describe the interfacial region [1]. Figure 1 gives a representation of a polyethylene crystal.

A. Crystalline Structure of Polyethylene

It is well recognized that polyethylene is a combination of crystalline and noncrystalline regions. The crystalline region is determined by the molecular structure (i.e., branching, degree of branching, types of branches, both inter- and intra-

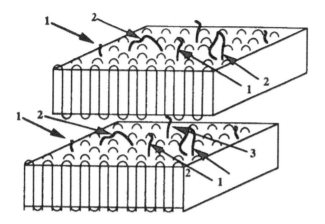

Figure 1 Schematic structure of polyethylene crystals: 1, cilia; 2, loops; 3, tie-molecule.

distribution of branches on the molecular scale), by the molecular weight, by the molecular weight distribution, and by the fabrication conditions. The combination of all of these factors controls the crystallization kinetics of the polymer, which in turn controls the morphology. A recent review article by Phillips [1] gives an excellent review of these effects. The crystallization process of polyethylene is a combination of nucleation (i) (deposition of the secondary nucleus on the crystal growth surface) and crystallite growth (g) (lateral spreading of the growing crystal on the crystal growth surface).

Variations between these two competing rates lead to three different crystalline growth mechanisms referred to as regime I, regime II, and regime III. When $i<<g$, the result is regime I, which corresponds to the classical picture in which the spreading rate is much greater than the nucleation rate. Regime II occurs when the two rates are comparable ($i \sim g$), and regime III results when the rate of nucleation is greater than the rate of spreading ($i>>g$). Since regime I has few nucleation sites relative to the spreading rate, crystals formed in regime I will have time for the molecules to reptate through the molten polymer and thus form large well-formed crystals. This will be the result when the melt of an unbranched polyethylene with a medium molecular weight is slowly cooled. As the molecular weight increases, or as imperfections on the backbone of the chain are added, the spreading rate is decreased. When g approaches i, the crystalline structure is regime II. Branched polyethylenes having low-density normally follow regime II, which results in smaller, less perfect crystals. Regime I polymers tend to have the molecules crystallize in the same crystallite, whereas the regime II polymers tend to have molecules incorporated into multiple crystals resulting in tie-molecules (see below). Cooling rate can also have an effect on the regimes and resulting morphology. Slow cooling, or annealing, favors regime I. Quench cooling, however, retards the ability of the molecules to reptate and thus favors regimes II and III. One can readily see then that quench cooling would lead to more tie-molecules than slow cooling and would tend to give tougher products. This is in fact what is observed.

Slow cooling will give a higher percent crystallinity, with well-formed crystallites, whereas fast cooling will give lower total percent crystallinity and smaller, less perfect crystallites. Annealing at temperatures slightly below the melting point will give the maximum crystallinity and largest crystallites. For high-density homopolymers the crystallization rate is so fast that it is difficult, if not impossible, to form spherulites. Therefore, the common crystalline structure is that of an axilite, or sheaf. The introduction of a very small amount of comonomer will affect the crystallization kinetics enough to result in the formation of spherulites, even when quenched cooled. At lower densities spherulites are not formed when the resin is quench cooled. Basically, all fabrication techniques for polyethylene quench-cool the fabricated article, with some being much faster than others (cast film is often extruded at >500°F and cast onto a chill roll at <100°F, whereas a thick-walled rotationally molded part may take several minutes to cool). Therefore, it is also obvious that fabrication conditions have a dramatic effect on the final morphology, and thus the properties, of the fabricated part.

There is general agreement that polyethylene crystalline regions are the result of chain folding. As the molecules fold and are incorporated into the crystal, they can either (1) immediately reenter the same crystal, termed adjacent reentry, which results in tight folds; (2) they can enter the same crystal at a nonadjacent location, which results in loops; (3) they can be incorporated into another crystal resulting in tie-molecules; or (4) they can terminate in the amorphous region to form cilia. Crystals grown from solution may have a large percent of adjacent reentry events, whereas crystals formed from quench cooling a melt will have more nonadjacent reentry events. Although there has been much discussion as to which mechanism is present, it is now generally agreed that both can exist depending upon the molecular structure and crystallization conditions. Phillips says, "The picture emerging of the molecular trajectory in rapidly cooled bulk systems is therefore one of the molecule remaining as a pseudorandom coil in which the molecule enters and re-enters crystals at a myriad of locations and does so in a way which always requires some limited form of adjacent re-entry folding, the remainder of the surface comprising largely tie-molecules, cilia and some non-adjacent re-entry folds" [1].

B.   Amorphous Structure of PE
Tie-Molecules

It has become increasingly apparent that to understand the failure mechanisms of polyethylene there must be an understanding of the noncrystalline region. Particularly important is the function of tie-molecules. Polyethylene crystallizes by a chain folding mechanism of polyethylene, as shown in Fig. 1. As a molecule becomes a part of a crystallite, one portion becomes anchored to the crystallite and the loose end is then "reeled in," or reptates, through the melt and is incorporated into the crystal. If for some reason a portion of the molecule is rejected from the crystal, it may then find its way into another crystal, make a loop with a molecule from another crystal, and then reenter the original crystal, or it may simply remain in the amorphous region. Molecules that are a part of two crystals or that form loops with molecules of other crystals then become tie-molecules, "tying" the two crystals together. Ends of molecules that reside in the amorphous domains may become entangled with other such molecules and form "weak" tie-molecules. The easier it is for a molecule to reptate through the polymer melt and become part of a crystal the fewer tie-molecules and the weaker the polymer. This is most likely to happen with low molecular weight high-density homopolymers. In fact, it has been shown that high-density polyethylene having a weight average molecular weight of about 20,000 or less has no measurable Izod impact strength.

Separation between adjacent crystallites, or lamellae, either through disentanglement or rupture of the tie-molecules, has been shown to be the dominant molecular mechanism in the environmental stress cracking of polyethylene [2–7], and in slow crack growth [8]. Tie-molecules have also been identified as exhibiting similar mechanisms in impact and tear strengths. Thus, tie-molecules are important to all strength properties of polyethylene. Brown et al. [8] have reviewed slow crack growth in polyethylene. They say that the basic failure process in slow crack growth involves the disentanglement of tie-molecules. The density of tie-molecules is the primary factor that determines the failure time. The most important quantity that distinguishes the different polyethylenes with respect to their resistance to long-time failure is the time for the beginning of brittle failure. For the same stress and temperature the time for initiating brittle failure can vary by a factor of 10^7 depending on the molecular and morphological structure of the polyethylene. They also believe that the molecular motion that controls slow crack growth is the same as that which occurs during the transition measured in the dynamic mechanical test.

Brown et al. [8] have shown that notch opening, and thus failure, is primarily dependent upon branch density. Homopolymer, with no branches, failed in less than 100 min, butene copolymer having 1.2 branches/1000 C failed in about 200 min, and butene copolymer having 2.0 branches/1000 C failed in more than 500 min, the length of the test. Copolymers with ethyl, butyl, and hexyl branches and about the same density of branches were also studied (note that this is the same density of branches, not the same density of the polymers). The activation energy was essentially independent of the length of the branches. The effect of branch density on the density of tie-molecules is related to the fact that the long period decreases as the branch density decreases. Table 1 shows the relationship between branch density and long period. In addition to increasing the number of tie-molecules by decreasing the thickness of the lamella crystals, the branches tend to inhibit the motion of the tie-molecules by pinning them. They note that the activation energy Q is about the same for the

Table 1 Relationship Between Branch Density and Length of Period

Branch density	T_m°C	Lamella thickness, L_c (nm)	Amorphous thickness, L_a (nm)	$L = 2L_c + L_a$ (nm)	Tie-Molecules fraction $t(\times 1000)$
0	133	33	13	78	1.64
0.08	131.5	27	15	70	9.2
1.2	130	24	15	62	15.4
2.0	129.5	23	16	61	38.3
2.3	128.5	21	15	57	54.7
4.6	126	17	16	50	92.7

homopolymer as for the copolymers. This observation suggests that the basic motion of disentanglement is the same. The difference is in the number of tie-molecules and the fact that many tie-molecules are pinned in the copolymer. The greater the number of tie-molecules, the smaller the disentangling force on each one, and thus the greater the strengh.

Molecules that are tied by crystals would be expected to resist being disentangled. If the crystals are very thin, they offer little resistance as the tie-molecules either move through the crystal or possibly shear it. If the lamella crystals are very thick, the number of tie-molecules is reduced. Probably the optimum occurs when the spacing between the branches is about equal to the thickness of the lamella, and the molecular weight is great enough to allow the molecules to be in more than one crystal.

The influence of molecular weight on the ESCR of PE has been shown by Haward et al. [9]. They related the molecular weight to post-yield strain hardening, and in turn related this to ESCR. A PE with $M_n = 1.11 \times 10^4$ and $M_w = 4.7 \times 10^4$ showed little post-yield strain hardening in a normal tensile test and exhibited a stress cracking time-to-failure t_f of 2 hr; a PE with $M_n = 1.55 \times 10^4$ and $M_w = 1.57 \times 10^5$ showed moderate strain hardening and a stress cracking time of 180 hr, whereas a PE with $M_n = 7 \times 10^5$ and $M_w = 6.4 \times 10^6$ showed strong strain hardening and a $t_f > 4000$ hr. The circumstantial evidence is that there is a relationship between the degree of strain hardening and ESCR. Since strain hardening is also a function of tie-molecule density, this is additional evidence that tie-molecules are important in the strength properties of PE.

Knowing these relationships will make it possible to design polymers having optimum strength properties.

C. Tie-Molecules in HDPE

The main cause for rejection of a polyethylene molecule from a crystal is the presence of imperfections, which are usually branch points, on the chain backbone. Without these branch points the molecules can easily crystallize and few tie-molecules will be formed, resulting in a polymer that lacks many strength properties. Since HDPE homopolymers have no branch points, it is necessary to find another way to increase the number of tie-molecules. The only molecular variables for HDPE are molecular weight and molecular weight distribution. Ergoz et al. [10] have shown that the molecular weight of HDPE has a large effect on the percent crystallinity. A HDPE with a $M_w = 1.15 \times 10^4$ has $(1 - \lambda) = 0.77$ (77% crystallinity), whereas a HDPE with $M_w = 1.2 \times 10^6$ has $(1 - \lambda) = 0.35$ (35% crystallinity). It has also been shown that the percent of interfacial content is inversely proportional to the molecular weight [11]. It ranged from ±5% at $M_w = 27.8 \times 10^3$ to ~12% at $M_w = 316 \times 10^3$. Thus, by increasing the molecular weight of HDPE two things are accomplished: (1) the density is reduced because a larger amount of the chains reside in the non-crystalline region, and (2) the larger molecules are "trapped" in multiple crystals or are highly entangled in the noncrystalline regions, thus increasing the number of tie-molecules, resulting in a "tougher" polymer. Regardless of which theory of crystalline growth is correct, it is evident that the crystallization process is dependent on the molecular weight.

IV. EFFECTS OF STRUCTURAL AND MORPHOLOGICAL FACTORS ON THE MECHANICAL PROPERTIES OF POLYETHYLENES

It is not practical to try to cover all combinations of molecular structure and processing conditions in this limited space. Therefore, attempts will be made to look at the various transitions that occur in the deformation of polyethylene under controlled conditions and then relate those to real-world applications. The most easily controlled test, and possibly the most reproducible, is the stress–strain tensile test. Since the ultimate mechanical properties of PE are determined by their deformation behavior, studies of this behavior under controlled conditions will lead to a better understanding of the relationships between molecular parameters and mechanical properties. Although the conditions of this test are not indicative of actual fabrication conditions, the test does serve to demonstrate the effect of various molecular and supermolecular structures on the deformation processes [12].

Figure 2 gives a "typical" stress-strain curve for polyethylene. Actually, there is no true "typical" stress-strain curve that applies to all polyethylenes, but this one is "typical" for slow-cooled linear polyethylenes and serves well to relate different structural properties to the deformation mechanisms. Figure 3 gives a series of stress–strain curves for various polyethylenes prepared at different conditions. It is readily apparent that the polymer structure and crystallization conditions have very significant effects on the deformation mechanism.

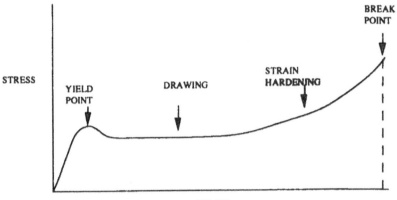

Figure 2 A stress-strain curve showing definitions of mechanical response features.

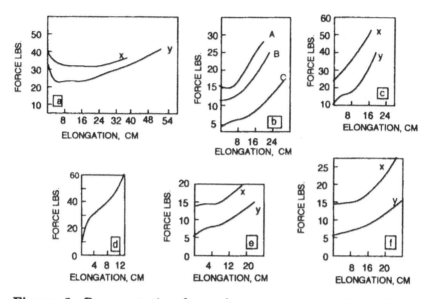

Figure 3 Representative force–elongation curves of polyethylenes. The force axis has been scaled so that all specimens have the same initial cross-sectional area. (a) Linear PE sample B: slow-cooled (x), quenched (y). (b) Linear PE sample C for different levels of crystallinity: A $(1 - \lambda)$ $\Delta H = 0.69$; B $(1 - \lambda)\Delta H = 0.59$; C $(1$ $\lambda)\Delta H$ = 0.47. (c) Linear PE sample D: slow-cooled (x), quenched (y). (d) Linear PE sample E slow-cooled. (e) Branched PE sample G: slow-cooled (x), quenched (y). (f) Ethylene–vinyl acetate copolymer sample G: slow-cooled (x), quenched (y). (From Ref. 12.)

These curves can be divided into three general categories. Samples in category 1 behave in the classical fashion [13,14]. They display neck formation at the earliest time of the draw. The deformation is confined to the region of neck formation, and the remainder of the sample shows no permanent deformation. The deformation is not homogeneous. Samples in category III show quite different behavior. There is no indication of neck formation in these samples at any time during drawing. The deformation is essentially uniform throughout the complete gage length for all drawing times. The deformation behavior of samples in category II is between the extremes of I and III. At low strain a neck is formed in a part of the specimen, but it is not sharp. After the neck forms, it propagates through the rest of the specimen. After that, the whole length of the sample deforms uniformly.

All of the tests were done at a 2 cm/min strain rate, therefore, the effect of deformation rate is not taken into consideration. Tables 2 to 4 summarize the samples used in this study [12]. The various stages of deformation, initial modulus, yield, drawing, and elongation, and tensile at break will be considered.

A. Initial Modulus

The initial portion of the deformation, up to about 3 to 5% strain, is usually reversible. The slope of the stress–strain curve at these low strain values is the modulus of the polymer, with 1% or 2% secant modulus, and Young's modulus being the most commonly used.

The initial modulus is a rather complex characteristic of a semicrystalline polymer such as polyethylene, because it is a function of molecular weight and both the crystalline and non-crystalline regions. Although there are some rules of thumb that apply to the modulus, exceptions to these occur due to other factors. For example, it is generally true that the initial modulus increases with density (or more accurately, degree of crystallinity); however, there can be significant exceptions to this caused by differences in molecular weight and/or crystalline structure [12,15]. There is also a direct relationship between initial modulus and crystallite size. It has been shown [12] that two polyethylenes having the same degree of crystallinity, (0.58 and 0.59), can have very different initial moduli (550 vs. 1000 MPa, respectively), which is related to the crystallite size, ~180 vs. ~290 Å. As might be expected, there is

Table 2 Molecular Characteristics of Polymers

Sample	Mol % short-chain branch	Mol % long-chain branch	$M_w \times 10^5$	M_w/M_n
A. Linear			0.53	2.90
B. Linear			1.5	12.5
C. Linear			5	5.5
D. Linear			20[a]	
E. Linear			80[a]	
F. Branched		0.61	1.5	9.6
G. Branched		1.06	3.5	18.5
H. Branched		1.58	9.5	57
I. Branched		1.29	2.2	20
J. Hydrogenated polybutadiene	2.20		1.08	1.3
K. Hydrogenated polybutadiene	2.20		4.20	2.66
L. Ethylene octene	1.37			
M. Ethylene-vinyl acetate[b]	1.43			
N. Ethylene-vinyl acetate[b]	2.20		0.83	5.3
O. Ethylene-vinyl acetate[b]	5.30			
P. Ethylene-vinyl acetate[b]	5.88		0.67	2.6
Q. Ethylene-methacrylic acid[c]	3.5			

[a]Viscosity average.
[b]Includes butyl groups.
[c]Includes ethyl and butyl groups.
Source: Ref. 12.

Table 3 Structural Characteristics and Mechanical Properties of Linear Polyethylene

Sample	Crystallization conditions[a]	$1 - l_d$ (%)	$1 - l_{DH}$ (%)	Morphology[b]	Crystallite thickness L_R (Å)	Yield stress (×10⁴ kPa)	Modulus (×10⁴ kPa)	l_B	Tensile stress at break (× 10⁴ kPa)
A	Sc	82	—	d	235	—	108.9	15.9	—
	Q (−78°C)	75	—	a	290	2.76	106.2	14.5	—
B	Sc, press	82	—	a	—	3.05	—	12.8	25.8 ±1.6
	Sc, air	77	—	a	235	2.91	115.5	—	—
	Q (−129°C)	65	—	a	150	1.93	101.4	13.4	39.4±2
C	121°/30 min	—	69	c	—	—	—	4.7	—
	105°/30 min	—	59	c	—	—	—	4.9	—
	Sc, press	70	—	c	—	2.67	108.9	—	—
	Sc, press	68	—	c	—	2.69	105.8	—	—
	Sc, air	66	—	c	275	2.45	93.8	6.3	36.9±1.1
	Q (85°C)	59	—	c	185	1.91	53.8	—	—
	Q (65°C)	56	—	h	165	1.85	51.4	—	—
	Q (−78°C)	53	47	h	—	1.66	48.9	5.2	—
	Q (−129°C)	52	—	h	160	1.58	46.9	6.0	27.6±4.2

D	121°/30 min	—	51	h	—	—	—	2.7	—
	105°/30 min	—	51	h	—	—	—	2.6	—
	Sc, air	61	—	—	—	—	—	3.2	13.6±1.1
	Sc, press	58	—	c	290	2.22	97.0	—	—
	Q (−78°C)	50	44	h	150	1.63	44.8	4.3	15.6±1.2
	Q (−129°C)	51	—	h	—	1.25	—	—	—
E	Sc	53	—	h	275	1.88	32.8	2.5	8.4 ±1.9
	Q (−78°C)	—	—	h	150	—	30.3	—	—

[a]Sc means slow-cooled, Q means quenched, temperatures and times are indicated

[b]Morphology: Letters a, c, d and h, refer to relative morphological structures where a has more perfect surpermolecular structures (spherulites) than b; d is lamellae organized into long thin rods; h is a random arrangement of the lamellar crystallites (no spherulites).

Source: Ref. 12.

Table 4 Structural Characteristics and Mechanical Properties of Structurally Irregular Polyethylenes

Sample	Crystallization conditions[a]	(Mol %) branches	$1 - l_d$ (%)	$1 - l_{DH}$ (%)	Morphology[a]	Crystallite thickness L_R(Å)	Yield stress (×10⁴ kPa)	Modulus (× 10⁴ kPa)	l_B	Tensile stress at break (×10⁴ kPa)
F	Sc	0.70	54	47	a	115	1.24	26.7	6.1±0.5	8.3±1.2
	Q(−78°C)	0.70	50	46	c	70	0.76	13.8	5.0 ±0.6	7.9±1.2
G	Sc	1.28	50	42	a	110	0.79	17.2	6.7±0.8	7.7±1.5
	Q(−78°C)	1.28	40	37	c	60	0.59	9.65	7.5 ±0.2	9.1±1.0
H	Sc	1.70	50	34	b	100	1.05	21.0	5.3±0.6	8.9±0.5
	Q(−78°C)	1.70	42	32	h	70	0.74	10.3	5.4±0.4	8.6±0.5
I	Sc	1.51	48	41	b	—	0.72	—	5.3±0.7	7.5±1.5
	Q(−128°C)	1.51	44	38	h	—	0.52	—	5.6±0.7	9.5±1.0
J	Sc	2.20	43	—	b	—	—	—	5.5±0.5	—
K	Sc	2.20	39	—	b	—	—	—	3.7±0.5	—

L	Q (−78°C)	1.37	—	—	b	—	0.68	—	6.4±0.4	11.8±2.0
M	Sc	1.43	—	36	c	—	1.29	33.2±5.4	5.5±0.3	7.2±1.0
	Q (−78°C)	1.43	—	36	h	—	0.96	13.9±1.0	5.5±0.6	7.8±2.9
N	Sc	2.20	—	32	a	72	1.04	21.7 Π±2.7	5.6±0.6	6.5±1.0
	Q (−78°C)	2.20	—	30	b	72	0.76	8.7±0.3	6.2±0.2	10.1±1.4
O	Sc	5.30	—	15	b	65	0.26	2.2±0.2	5.4±0.3	14.0±2.4
	Q (−78°C)	5.30	—	17	h	57	0.29	2.0±0.1	5.3±0.2	13.2±2.5
P	Sc	5.88	—	14	non-lamellar	70	—	1.2±0.1	4.9±0.2	—
	Q (−78°C)	5.88	—	15	non-lamellar	—	—	1.3±0.1	4.9±0.2	—
Q	Q (−78°C)	3.50	—	25	—	—	0.40	—	4.3±0.35	—

aLetters have some meaning as in Table 3.

Source: Ref. 12

also an inverse relationship between the thickness of the noncrystalline region and the modulus. However, for a high molecular weight polymer, $M_w = 8 \times 10^6$, there is no change in modulus even with a twofold change in crystallite thickness. It has recently been found that there is also a difference in moduli between two polymers having the same melt index (MI) and density, but having different branching distributions. A polymer of 1 MI, density of 0.902 g/cc, and homogeneous branching distribution had a Young's modulus of 6464 psi. whereas a polymer of 0.8 MI, density of 0.905 g/cc, and heterogeneous branching distribution had a modulus of 8144 psi. One mil films made from these two resins had 1% secant moduli of 8035 psi and 11790 psi, respectively (both of these are machine direction (MD) values). One possible advantage of the effect of homogenous branching distribution having a lower modulus than a corresponding heterogeneous polymer would be that one could obtain the same modulus with lower amounts of comonomer. This is yet to be shown experimentally.

From these relationships one would expect an increase in modulus with (1) an increase in density, (2) a medium molecular weight, (3) a large crystallite size, and (4) a heterogeneous branching distribution, which may be effective because it tends to give a distribution of crystallite sizes, with many of them being very large, whereas homogeneously branched polymers tend to give very uniform and relatively small crystallites.

B. Yield

The first nonrecoverable strain in the deformation of polyethylene is what is commonly referred to as the yield point or yield stress. To a first approximation the yield stress looks like it is a linear function of the crystallinity. Comparisons made between polymers having the same crystallinity but different supermolecular structure show that they have about the same yield stress; thus, the contribution of different crystalline structures (spherulities vs. axilites, or different sizes of crystallites) on yield stress appears to be minimal (Fig. 4). However, Fig. 5

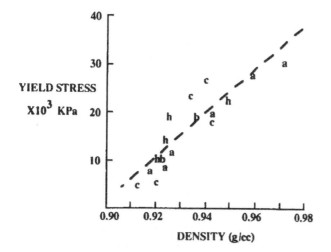

Figure 4 Yield stress as a function of density. a, b, c, and h have the same meaning as in Table 3. (From Ref. 12.)

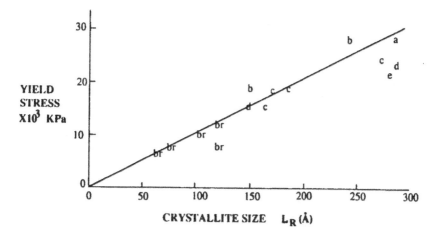

Figure 5 Yield stress as a function of crystallite size for linear and branched polyethylenes. a, b, c, d, and e represent linear PE samples of the same (capital) letter and br represents branched PEs. (From Ref. 12.)

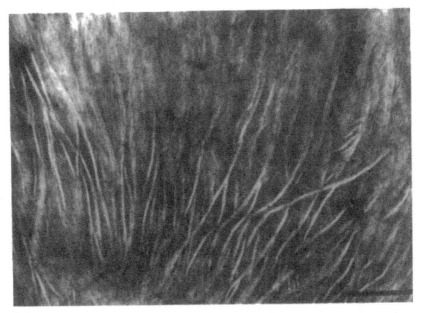

(a)

Figure 6   Electron micrographs of LLDPE before drawing: (a) ring region; (b) center region of spherulite. Scale bar = 0.5 μm. (From Ref. 15.)

does show a relationship between the yield stress and the most probable crystallite thickness, $L_R{}^*$. Not only does this plot show a relationship between yield stress and crystallite thickness, but it extrapolates to the origin at zero thickness, which corresponds to the absence of crystallinity. Therefore, a completely amorphous polymer above its glass transition temperature would exhibit rubberlike elasticity and have no yield point.

The mechanism, or mechanisms, for permanent deformation of polyethylene is also open to some debate. Again, probably more than one mechanism contributes, depending upon the polymer structure and the test conditions. Furuta and Kojima [16] have done an excellent study of the morphology of the

(b)

deformation process of linear polyethylene. They have shown, via electron microscopy, the different morphological changes that occur at the different stages of deformation. Figure 6 shows the morphology of the spherulites prior to drawing. At a draw ratio of 3, which is past the yield stress and into the early stages of drawing, they observed that the changes of morphology depend on the location in the spherulite, (Fig. 7). They showed that the periodically rotated lamellae disappear and the lamellae are aligned parallel to the draw direction in the polar regions (region A) and are separated and more or less perpendicular to the drawing direction in the equatorial zone (region B) in the spherulite. The entangled and twisted lamellae that were in the center of the spherulite (Fig. 6) are not seen in the figure. The stress apparently resulted in their disentanglement at an early stage of drawing, possibly at the yield point.

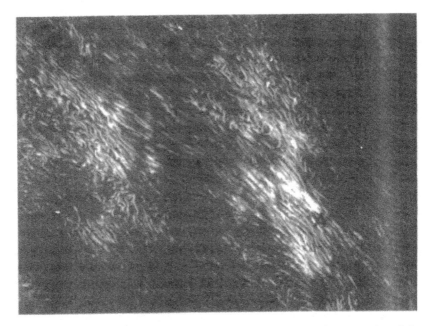

Figure 7   Electron micrographs of LLDPE for a draw ratio of 3. Deformed spherulite at low magnification. Regions A and B represent polar and equatorial regions respectively. (From Ref. 15.)

Popli and Mandelkern [12], as well as others, believe that the molecular mechanism of deformation is that of partial melting and recrystallization. It is understood that there is adiabatic heating during the deformation [17–20]. They point out that when the thermal process is coupled with the applied stress, partial melting and recrystallization should occur. The partial melting involves the fusion of less perfect crystallites, which will then recrystallize in the draw direction. The recrystallization of an oriented or partially oriented amorphous system will result in a decrease in stress. Flory and Yoon [21] and Hookway [20] have also shown that raising the temperature is not a necessary condition for partial melting to take place, and that the applied stress is adequate by itself. A decrease in the nominal stress is a major characteristic of the yield point. For those systems that deform inhomogeneously (with necking), some reduction in the nominal stress is to be expected because of the reduced cross section of the specimen. However, the maximum cannot be attributed to the geometric changes alone [22]. Therefore, a substantive molecular contribution to the maximum is observed at the yield point. This can also explain the relationship of yield stress versus L_R^* in that the thicker crystallites are more thermally stable and thus require more stress to induce the partial melting and recrystallization. These deformation processes probably are dominant throughout the drawing region.

C. Draw Area Deformation

Following the yield stress, in those polymers that show a maximum yield stress, there is a drop in the stress. In polymers that do not show a maximum at the yield stress there is usually a leveling off of stress. In each case there is a stress plateau where there is little or no increase in stress with an increase in strain. The deformation mechanism(s) during this region are assumed to be the same as in the initial yield, but exert their influence to a greater extent. The maximum draw ratio at break is a function of both the average density of the polymer and the structures of the molecules that make up the final polymer. High-density polyethylene will give a higher degree of elongation at break than will a polyethylene with a low density, both LDPE and LLDPE. Heterogeneous LLDPEs at low densities, 0.90 to 0.92, appear to have higher percent elongation than homogeneous LLDPE by about 100 to 200% (800% vs. 600%), but at higher densities they are about the same. This may be caused because the presence of the "high-density" fraction in the heterogeneous resins has a greater effect at the very low densities. The data for these comparisons are sketchy at this time and additional data are needed to confirm these initial observations.

D.   Strain Hardening

After the polymer elongates to some value, depending on the particular polymer, there will again be an increase in the stress with an increase in the strain. This region is referred to as strain hardening. In this process imperfectly oriented chains are brought into better alignment by the increasing stress, which requires a large input of energy. The degree of strain or percent elongation at which this strain hardening begins is dependent primarily upon the density. However, the main effect of density is at very high densities, about 0.940 and above. Below this value there is not very much difference in the percent elongation with a change in density. From about 0.940 density to a homopolymer (all of the same molecular weight) the percent elongation ranges from about 1300 to about 1500, and from 0.90 to about 0.93 it ranges from about 600 to about 800.

E.   Draw Ratio at Break

After strain hardening begins, the stress will rise with increased strain until rupture of the specimen occurs, giving the ultimate strength of the polymer. Popli and Mandelkern [12] have also studied the polymer characteristics that affect the ultimate elongation, or elongation at break. They found that the relationships differ depending on whether or not the polymers were linear (unbranched) or branched. Their studies included free-radical polymerized homopolymers, hydrogenated polybutene, one octene copolymer, and ethylene–vinyl acetate and ethylene–methacrylic acid copolymers. They show that there is a distinct difference between the draw ratio at break of the linear and nonlinear polymers. The draw ratio at break for the linear polymers ranged from about 6 to about 16 and was correlatable with the M_w, with the draw ratio at break decreasing as the molecular weight increased (Fig. 8). Whereas the draw ratio at break for the branched polymers was only

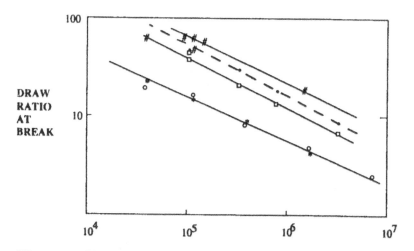

Figure 8 The maximum draw ratio available as a function of molecular weight: *, linear PE, slow-cooled; ○, linear PE, quenched; #, linear PE at 75%°C; □, HMW HDPE at 75°C; ●, HMW HDPW at 95°C. (From Ref. 12.)

about 4 to about 7 and showed no correlation with M_w. There was not enough data on linear low-density copolymers to draw any conclusions. However, there are indications that a relationship exists between both the branch content and molecular weight for fractions of LLDPE and the draw ratio at break.

F. Ultimate Tensile Stress

The ultimate stress, or break stress, is a function of the molecular weight and crystalline structure. Popli and Mandelkern [12] have shown that the ultimate strength for linear, or high-density, polyethylene is inversely proportional to the molecular weight in the range of 10^5 to 10^7 M_w. At 10^5 M_w the break strength was about 400 MPa and at 10^7 M_w it was about 100 MPa. For copolymers of ethylene and low-density polyethylene there was a decrease in ultimate tensile strength with a

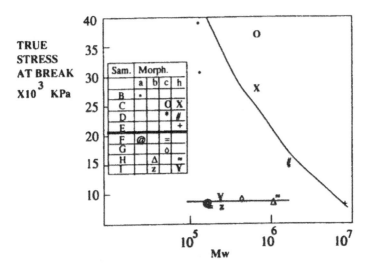

Figure 9 Ultimate strength as a function of molecular weight. Samples B to E are linear PE. Samples F to I are branched PE. (From Ref. 12.)

decrease in density up to about 2 mol % of branch points where it leveled off at about 100 MPa (Figs. 9 and 10).

G. Summary of Stress–Strain Curves

The importance of the stress–strain deformation of polyethylene is that it gives some understanding that different mechanisms occur during the deformation of polyethylene, although we may not fully understand what the mechanisms are. They also show that during the deformation different energy levels and total energy are required to carry out the different deformations, for example, the maximum stress before yielding, the yield stress, the stress required to draw the polymer after yielding, and the stress required to break the sample. This combination of stress and strain, or area under the stress–

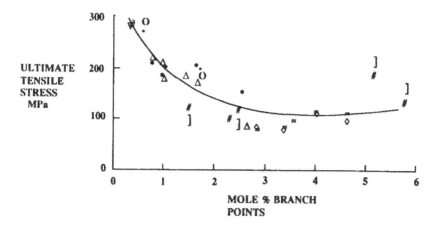

Figure 10 Ultimate strength as a function of degree of branching: ●, ethylene butene quenched; O, ethylene-butene slow-cooled; Δ, ethylene-octene slow-cooled; *, ethylene-octene quenched;], ethylene–vinyl acetate slow-cooled; #, ethylene–vinyl acetate quenched; ▽, ethylene-hexene quenched; ▽, ethylene-hexene slow-cooled; ≈, ethylene–methacrylic acid quenched; ◇, ethylene–methacrylic acid slow cooled. (From Ref. 15.)

strain curve, is related to the total "toughness" of the polymer and is often called the polymer toughness.

The major characteristics of deformation curves allow them to be divided into three categories (Fig. 11):

Category 1 (samples J and N): Polymers that have about 10% crystallinity or less have little or no supermolecular structure. Some have suggested that they do not contain lamella crystallites [23]. These polymers can be considered as consisting of very small crystallites dispersed in a matrix of disordered chains.

    Because of the broad dispersion of the very small crystallites, the noncrystalline chains have little restraint in their movements. When stress is initially applied to such

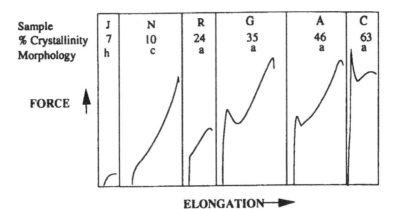

Sample
% Crystallinity
Morphology

FORCE

ELONGATION

Figure 11 Stress–strain curves for polyethylene copolymers: J, ethylene-octene copolymer, slow-cooled; N, ethylene–vinyl acetate copolymer, icewater quenched; R, ethylene–methacrylic acid copolymer, slow-cooled; G, ethylene-octene copolymer, slow-cooled; A, ethylene-butene copolymer, slow-cooled; C, ethylene-hexene copolymer, slow-cooled; Morphologies are as in Table 3. (From Ref. 15.)

polymers, it mainly deforms the noncrystalline regions in a recoverable manner, bypassing the crystallites, thus resulting in low modulus. As the stress is increased, the crystallites become involved, but because of the random nature of the dispersion of the crystallites some will be more stressed than others, resulting in preferential deformation, allowing crystallites to deform almost independently of each other. This results in homogeneous deformation without necking or a yield maximum.

Category 2 (samples R and G): As the level of crystallinity increases, lamellae form and the amount of noncrystalline material between the crystallites decreases. It is no longer possible for the noncrystalline regions to deform independently of the crystalline regions. This gives rise to an increase in the initial modulus as the crystallinity develops. The crystallites are constrained, compared to the previous polymers where the crystallites are free to act independently, and thus must deform over a complete cross section. This results in necking, which in turn results in a maximum yield point, or stress, which is broad.

Category 3 (samples A and C): This category consists of polymers with the highest crystallinity and well-developed superstructures. Here the interlamellar region, which contains the noncrystalline components, is relatively small and cannot be deformed independently of the crystallites. Thus, the crystalline and noncrystalline regions will be deformed in parallel, and the value of the modulus will be an approximate linear function of crystallinity. These polymers show a sharp yield maximum and a well-defined neck.

Data that have been obtained [12,15] indicate that the factors determining the initial modulus are a combination of crystallinity level, the lamellar structure, the interlamellar thickness and, in some respects, the supermolecular structure. As would be expected, these characteristics are controlled by the molecular weight, the concentration of branches, and the crystallization conditions.

Peacock and Mandelkern [15] concluded that for whole copolymers the tensile properties, up to and including the yield point are controlled primarily by the degree of crystallinity or, conversely, by the noncrystalline portion. There is little or no specific dependence on branch type, concentration, and supermolecular structure. After yield and neck propagation have taken place, the ultimate properties do not show a significant dependence on the molecular constitution, phase structure, or morphological features. The samples they used were not characterized by molecular weight or comonomer distributions. We have found that both affect the stress–strain relationships for LLDPEs.

V. TOUGHENING OF POLYETHYLENE

Now that we have some idea of what contributes to the strength, or toughness, of polyethylene, how can it be made "tougher"? Some commonly used methods are applicable to all types of PE and for all types of applications. These will be discussed in conjunction with the methods used, the types of polymers, and the applications.

A. High-Density Polyethylene (HDPE)

Blown films made from high-density homopolymers tend to be very splitty in the machine direction, because of the lack of tie-molecules and the fact that the HDPE tends to be oriented in the machine direction. High-density polyethylene is used extensively in applications where there is a need for improved barrier properties, high rigidity, or high yield strength. Since HDPE tends to have few tie-molecules, how can its strength be improved without sacrificing its desired properties? Four ways are commonly used to make tough high-density polyethylene: (1) increasing the molecular weight, (2) incorporating small amounts of comonomers, such as butene, hexene, or octene, (3) crosslinking, either chemically or with radiation, and (4) blending with "impact modifiers." *Toughness* refers to any physical property that is required for the polymer to have an acceptable performance and includes one or more of the following: impact strength, tear strength, environmental stress crack resistance, slow crack resistance, yield strength, and modulus. Toughness is also defined as the area under the stress–strain curve, as discussed earlier.

1. EFFECT OF MOLECULAR WEIGHT

If the application requires very high modulus, such as for large bottles or drums, a homopolymer, or copolymer with a density higher than 0.95, is usually required. In order to build "toughness" into a HDPE homopolymer it is necessary to make a high molecular weight polymer. The molecular weight will depend on the particular application. For example, milk bottles must be very rigid, and since they have very thin walls HDPE is required. However, since the application is not too demanding, because of the relatively low strength requirements, low ESCR requirements, and very short time use (in fact, milk bottles may have the shortest time use for any bottle), a rather low molecular weight of about 125,000 M_w (melt index = 1) is sufficient. Drums and large containers made via blow molding are usually used in very demanding service, such as containers for chemicals, and therefore are required to have high modulus, excellent ESCR, high impact strength, and fatigue resistance. To accomplish this feat the resins are very high in molecular weight, usually over 250,000 (~0.01 melt index).

Some HDPE film resins must have a very high molecular weight to prevent splittiness. These resins may have a melt index of less than 0.01. The very high molecular weight molecules have a much more difficult time reptating through the melt and thus end up being incorporated into more than one crystallite or forming loops with other large molecules, and therefore become tie-molecules. These make very tough, high-modulus films that are used for applications all the way from newspaper bags to grocery bags to pond liners. Because of the combination of toughness and modulus they can be drawn into very thin films and still have excellent properties.

2. EFFECT OF COMONOMER

A second way to improve the "toughness" of HDPE is to copolymerize a small amount of an α-olefin during the polymerization of the ethylene. The addition of a comonomer that reduces the density, an indication that an increasing amount of the polymer is in the amorphous state, can result in a significant improvement in strength properties, but at the sacrifice of properties dependent on modulus. HDPE is defined as having a density of 0.945 and above, with the desired density varying depending on the application. Most nonfilm applications require a higher modulus than is given by a 0.945 density and usually require at least 0.950. The main applications for these types of polymers are for large containers: bottles, drums, and various types of tanks and pipe. These applications require a balance of modulus, ESCR, creep resistance, and overall toughness.

The most common comonomer used for improving the toughness of HDPE is butene. Butene copolymers are commonly used for applications that require high ESCR but that are of relatively short duration, such as detergent bottles. However, for applications that demand high ESCR and have a long life, such as pipe, a higher α-olefin, such as hexene, is used. Phillips Petroleum has used hexene copolymers for making pipes with superior ESCR and hoop stress strength since the early 1960s.

The comonomers serve as defects along the polymer chain, interfering with crystallization and resulting in tie-molecules which greatly improve the physical properties. Studies of molecular weight fractions of various HDPEs have shown that polymers that have the branching in the high molecular weight fraction have improved ESCR. This would be expected, since the large molecules would be the ones most capable of being tie-molecules, which would be enhanced by branching.

3. EFFECT OF CROSSLINKING

A third common method of improving the toughness of HDPE is by crosslinking, and the most common crosslinking agents are peroxides, although silanes and radiation are also used. Crosslinking agents cause the formation of branching and networks in the polymer. Both of these structures lower the mobility of the molecules in the melt, reduce the degree of crystallinity (lower density), and increase the number of tie-molecules, resulting in improved toughness. Large agricultural tanks are a prime example of a very demanding application of HDPE. These tanks must be high density to have adequate modulus to support the large volumes, up to 5000 gallons, and must also withstand aggressive environments caused by some agricultural chemicals. Most, if not all, of these large tanks are molded via rotational molding using a peroxide for crosslinking. Since only the weight of the polymer causes it to fuse and flow in rotational molding, there is a limit to how high the molecular weight of the polymer can be. However, by adjusting the amount of peroxide and cure conditions, it is possible to make tanks with high modulus that are highly crosslinked. These tanks have high resistance to agricultural chemicals and weather conditions (e.g., heat and ultraviolet light). Phillips Petroleum sells resins made using their slurry process that already contain peroxide for these applications. Converters may also add peroxide to fit their particular needs.

4. EFFECT OF BLENDING WITH IMPACT MODIFIERS

A fourth, and less common, method for toughening HDPE is to add some type of rubber. This will improve the "toughness" of the resin, particularly when it is converted into film, but tends to destroy some of the desirable properties of HDPE. Addition of rubber has a marked effect on the modulus, which decreases as the amount of rubber is increased, and on barrier properties, which are also diminished as the amount of rubber is increased. The same results, loss of modulus and barrier properties, are obtained by adding other polyethylenes such as LDPE and LLDPE, although these will improve some of the film properties.

B. Low-Density Polyethylene (LDPE)

Low-density polyethylene is produced by polymerizing ethylene from about 16,000 psi up to about 40,000 psi using a free-radical initiator such as peroxide. The type of reactor dictates the pressure range used (autoclave reactors normally range from about 16,000 psi to about 26,000 psi, and tubular reactors normally are 30,000 psi and up). No details of the reactors will be included in this discussion; they are mentioned only because different types of reactors give different types of polyethylene, even though they are all LDPEs. Because of the free-radical polymerization mechanism, all low-density polyethylenes, including copolymers made via the high-pressure processes, have a mixture of short-chain branching (branches less than six carbon atoms long) and long-chain branching (the length of these branches cannot be determined, but branches are generally long enough to have a measurable effect on melt properties). There is some evidence that a C_{18} chain begins to have a slight effect on some of the melt properties, but it is generally believed that the long branches in LDPE are longer than 18 C's. The density range for LDPEs is about 0.915 to 0.932 for the base resin, but owing to additives, the final densities are often higher than the base density. LDPE is used primarily for making film, and thus the main emphasis on the toughness of the polymer is related to film properties, namely, impact strength and tear strength. ESCR is important where seals may be subjected to aggressive environments, such as in bags used to contain food. Unlike HDPE, optical properties (clarity, gloss, and haze) are often the most critical properties for a given application, even at the expense of strength properties. There are three main ways of toughening LDPE (which will be discussed in the following sections), or four ways if one includes making copolymers with non-hydrocarbon comonomers such as vinyl acetate or acrylic acid. This discussion will be limited to homopolymers.

As in any polyethylene the first way considered to toughen LDPE is to increase its molecular weight. One advantage of LDPE over HDPE is that the long-chain branching in LDPE causes it to be more shear sensitive in the melt, therefore it is easier to process at a given molecular weight. Most film applications requiring a "tough" LDPE can be satisfied with a resin having a MI = ~0.1-0.2, which is the limit of the molecular weight that can be economically achieved in the high-pressure process. The most demanding high-volume applications, such as heavy-duty bags, are usually satisfied by these polymers. Crosslinking of LDPE is also a common way to improve its "toughness"; however, crosslinking is often used to impart heat-shrink properties and not for the purpose of toughening the film. Crosslinked polyethylene has improved ESCR, enhanced tensile and impact strength, and superior dimensional stability and it can be used at higher operating temperatures. The two common methods for crosslinking polyethylene use either radiation (gamma or beta) or the addition of organic peroxides. Silanes are also used. The most widely used application of crosslinked LDPE, in which the purpose of the crosslinking is to impart toughness, is in the wire and cable industry. Here LDPE is crosslinked primarily to inhibit creep and "treeing" in power cables, or other very demanding applications. The starting material will have a melt index of about 2 or greater so that the peroxide can be added below its initiation point; then the crosslinking will be carried out above the degradation temperature of the peroxide. The level of crosslinking is controlled by controlling the amount of peroxide, usually ranging from 0.1 to 2%. Silanes are also used in this application and have the advantage of not being as temperature sensitive as the peroxides, but they are activated by passing the wire through a water bath to cause crosslinking.

Irradiation crosslinking is done quite extensively to make shrink film. In this process a tube of polyethylene is extruded, cooled, and then irradiated, using beta radiation. Gamma radiation is often used for sterilizing polyethylene film and parts, but the dose rate is too small to be used in this application. After irradiation, the tube is heated and postinflated to the desired size and cooled. This type of film is used for making bags or wrapped film that when heated to its softening point will shrink very tightly around the contents. The crosslinking not only imparts the shrinking characteristics to the film, but also makes the film tougher.

Matusevich et al. [24] have compared the efficiency for peroxide versus irradiation crosslinking. They irradiated LDPE at different doses and crosslinked the same LDPEs using different levels of dicumyl peroxide. They then measured the strain versus stress for the different levels of crosslinking (Fig. 12).

Based on their data they derived the following relationship between peroxide and radiation crosslinking:

$$C_{DCP} = 8.3 \times 10^{-6}D - 1.3$$

C_{DCP} is concentration of dicumyl and D is the amount of gamma radiation in Gy.

Silane crosslinking can be accomplished by adding trialkoxy-vinylsilane in the presence of a radical initiator (e.g., a

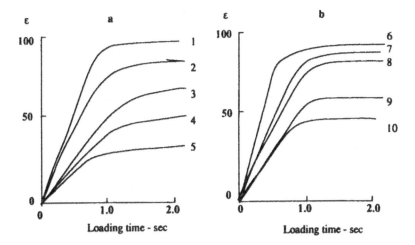

Figure 12 Strain versus stress for various levels of crosslinking. (a) Radiation in Gy: 1, 5×10^4; 2, 1×10^5, 3, 2×10^5; 4, 5×10^5; 5, 1×10^6; (b) Wt % dicumyl peroxide: 6, 0; 7, 0.3; 8, 0.5; 9, 1.0; 10, 2.0; ϵ = % strain (From Ref. 23.)

peroxide) to graft the silane onto the polyethylene. The cross-linking is achieved by the exposure of the grafted polyethylene to water resulting in the hydrolysis and subsequent condensation of the grafted alkoxysilane groups.

The third common way of toughening LDPE is to blend a rubber, or more commonly today, linear low-density polyethylene (LLDPE), into the LDPE. The addition of LLDPE to LDPE is the most common method used for improving the strength of LDPE used in blown and cast films and extrusion coating, as well as in molding applications. The critical stress intensity factor, K_c, which can be related to "toughness," is from 12 to 35 times as high for LLDPE as for LDPE [25]. This is due to LLDPE having more tie-molecules than LDPE and indicates that the addition of LLDPE to LDPE should result in increased toughness of LDPE. Since the commercialization of LLDPE in 1978 there have been many literature references to such blends. It is to be expected that there are even more applications than have been reported. When one considers the number of LDPE and LLDPE products, and the possible blend ratios, it is obvious that the number of possible blends is almost infinite. It would be futile to attempt to give meaningful examples in this brief summary; therefore, only general observations will be made.

Because LLDPE contains no long-chain branching it will be harder to process, have poorer bubble stability, and show more neck-in (for cast film or extrusion coating) than LDPE (it does draw down more easily than LDPE). Therefore, these factors must all be considered when making blends. Maxwell et al. [26] have shown that the blending conditions can have an effect on the properties of blends of LDPE and HDPE and they conclude, "It is possible to produce blends with either greater or less melt elasticity than the individual components in the blends. This demonstrates that the correlation of properties of blends with the properties of the components should not be attempted without consideration of the effect of the blending process on the properties of the components themselves."

Although Maxwell et al. used HDPE, it is reasonable to believe that the same precautions should be used when using
LLDPE.

Generally, the addition of LLDPE to LDPE will result in a
blend with the following properties:

1. *Poorer processability:* higher pressures, and or temperatures; poorer bubble stability; more neck-in; better draw-
   down; lower melt elasticity (melt tension).
2. *Tougher products:* greater film "toughness" (impact
   strength, tear strength, puncture resistance, and ESCR) for
   blown, cast, and extrusion-coated film; improved toughness, ESCR, impact strength for molded parts.

It is often necessary to compromise between the improvement
in physical properties and the sacrifice in processability of the
blend.

A large number of basic studies of blends of LLDPE and
LDPE were done in the early 1980s, both inside and outside
Dow. These include fundamental studies of blends [27–33],
blends for films, injection molding, drip irrigation tubing, extrusion coating, and blends of LLDPEs with elastomers for
films and injection molding.

C. Linear Low-Density Polyethylene
(LLDPE)

Linear low-density polyethylene is a class of polyethylenes
that are copolymers of ethylene and alpha olefins, usually C_4
to C_8. Although DuPont supplied various LLDPE copolymers
in the 1960s under the trade name Sclair®, they did not become a major product industry wide until the late 1970s when
Dow developed Dowlex® and Union Carbide commercialized
their gas-phase process for making LLDPE. The name linear
low density was given to these copolymers because they do not
have long-chain branches like LDPE, and thus process like

linear HDPE, but they do have low densities and thus have many properties like LDPE.

The linear low-density resins that are copolymers of ethylene and 1-butene generally have better strength properties (tear and impact) than do LDPEs at the same molecular weight (MI) and percent crystallinity (density). However, they tend to have poorer optical properties (in blown film) and are much harder to process. The difficulty in processing is due to narrow molecular weight distribution (MWD) and lack of long-chain branching. Data have shown that at the same MI and density copolymers of higher α-olefins ($>C_4$) have better strength properties than do C_3–C_4 copolymers. One question that remains to be answered is, How much of this effect is due to the length of the side branch and how much is due to other factors, such as short-chain branching distribution (SCBD), caused by the difference in reactivity ratios of the different monomers and resulting in differences in SCBD? Copolymers made by different processes, such as gas phase versus solution, have significant differences in SCBD, and sometimes in MWD. Since all of these factors (MW, MWD, SCBD) affect physical properties, it is not possible to assign quantitative values to the effect of each.

The structural differences between LLDPEs and LDPEs are (1) the lack of long-chain branching (LCB) in the LLDPEs and (2) uniform length of the short branches in LLDPEs. Most of the branches in LDPE are either ethyl or butyl, with the ratio of butyl/ethyl branches decreasing as the total number of branches increases [35]. Since the length of the branches in LDPE are the same as for butene-1 and hexene-1 ethylene copolymers, one might anticipate that the physical properties would be about the same. However, it is well documented in the literature that LDPE has higher melt shear sensitivity and higher melt elasticity than do the LLDPEs. Also, it is well documented that at the same molecular weight and density LLDPEs, both butene-1 and higher alpha-olefin copolymers, have higher impact strengths, greater tear strengths, and better environmental stress crack resistance (ESCR) than LDPE. It is rather easy to assign the differences in melt properties to the presence of LCB in LDPE, because the LCB would tend to increase the entanglement at low shear rates (high elasticity), but at high shear rates the chains would disentangle, thus reducing the viscosity. This would result in more shear sensitivity. In the solid state the LCBs apparently interfere with the formation of tie-molecules and crystalline growth (making smaller and thinner crystallites, possibly by interfering with reptation), which results in more and weaker crystal interfaces.

The same methods used to "toughen" HDPE and LDPE are used to toughen LLDPE. However, LLDPE is generally much tougher than comparable HDPE and LDPE and is modified for only the most demanding applications, such as wire and cable applications where it may be crosslinked. Many applications that at one time used modified LDPE or HDPE, such as heavy-duty bags, can use LLDPE without modification.

D. Polyethylene Composites

Addition of fillers to polyethylene is a common practice in molding applications and some extrusion applications. However, it is not common in film, except in fairly low concentrations of only a few percent to perform as a pigment, the most common being titanium dioxide.

The fillers most commonly used in polyethylene are calcium carbonate, talc, kaolin, mica, silica, and fiberglass. Fillers are generally added to polyethylene to increase the modulus, abrasion resistance, and heat distortion temperature, often at the expense of toughness or impact strength. They are especially used in larger injection-molded or blow-molded articles. As in other polymer systems, the types and form of the fillers markedly influence their effect on the properties of the polyethylene. The usual important characteristics of the filler are the morphology and surface activity.

Since polyethylene is very nonpolar it does not tend to bond chemically to any fillers. Thus, fillers may cause the polymer to be stiffer, but at the same time it will decrease in toughness, impact strength, and ultimate stress elongation. The size of the particle, its dispersion within the polymer matrix, and the interfacial adhesion of the particle to the polymer matrix are very important to the properties of the composite. Small particles that are well dispersed generally give better properties. However, it is often difficult to disperse very fine particles because of their tendency to agglomerate. To overcome this problem it is common to use fillers that have been "treated" with a surface agent that permanently modifies the filler's organophilic nature and makes it more easily dispersible. Typical surface-treating agents include stearic acid, calcium stearate, amines, and quartenary ammonium compounds.

Coupling agents, another class of compounds, aid in making the fillers more compatible with the polymer and thus improve their properties. These not only act as surface treatment agents, enhancing dispersability, but they promote strong interfacial adhesion between the particle surface and the polymer. The most common coupling agents used in polyethylene are organic silanes and organic titanates. These apparently have similar functions and effects on the properties of the composites. Recently, Malik [35] has shown, via electron microscopy, that mica treated with silane had stronger adhesion to HDPE than mica treated with titanate. The HDPE containing the mica treated with silane also had better physical properties than HDPE containing mica treated with titanate.

A very brief summary of the effect of fillers on properties of polyethylene follows:

Modulus: Modulus is the property that is most enhanced by the addition of fillers. The maximum increase in modulus with the minimum detrimental effects on other properties is obtained by using a filler with fine particle size and a coupling agent.

Tensile strength: Tensile strength can be either enhanced or decreased by the addition of fillers. Fortunately, the same filler characteristics that help modulus also improve tensile strength, that is, fine, well dispersed, and compatible (surface-treated or containing a coupling agent).

Elongation: Generally, as the level of filler increases the elongation decreases. Exceptions can occur where the organic groups of a coupling agent simply entangle with the molecules of the polymer instead of crosslinking with them. In such cases tensile strength can be improved without severely affecting the ultimate elongation. However, if there is crosslinking between the coupling agent and the polymer, the elongation will decrease.

Impact strength: Addition of hard, inorganic fillers to polyethylene will decrease impact strength unless there is good interfacial adhesion between the filler and the polymer matrix. This may require more than just coupling agents. Copolymers of ethylene that are polar (such as ethylene acrylic acid or methacrylic acid copolymers) can strongly bond with fillers and at the same time be compatible enough with the polyethylene matrix that the impact strength is maintained.

Tear strength: The characteristics of fillers and polymers that affect impact strength have a very similar effect on tear strength. In a recent European patent application Robinson et al. [34] have disclosed a unique way of toughening LDPE. They blended "high performance polyethylene powder" into LDPE and a 50/50 blend of LDPE and LLDPE. The "high performance powder" was described as "finely divided fibers of a linear polyethylene having a molecular structure comprising substantially parallel macromolecules." These types of fibers are produced by drawing fibers from a "gel" that is made by dissolving ultrahigh molecular weight HDPE in a solvent and cooling. The resulting fibers are highly oriented, have extremely high modulus in the draw direction, and have a melting point above 140°C. A blend that contained 100 parts LDPE, 40 parts of carbon black, and additives was exposed to 15 Mrad of radiation. The resulting blend had a tensile strength of 17 MPa, an elongation of 600%, and a 2% secant modulus of 198 MPa. A similar blend of 100 parts of LDPE and 20 parts of the "high performance powder" had a tensile strength of 45 MPa, elongation of 800%, and a 2% secant modulus of 420 MPa. Since the "high performance powder" sells for about $20/lb, this is not practical, but it is interesting.

VI. SUMMARY

Polyethylene is the highest-volume polymer sold in the world today. This high usage is due to its extreme versatility. Most of its desirable properties are inherent within the polymer, but it also lends itself to several different avenues of modification to improve its properties. Such modifications are usually done to make it even tougher and to give it higher modulus. This review has given an overview of the inherent properties of polyethylene as well as some of the more common methods of improving some of the properties of polyethylene, especially toughness and modulus.

REFERENCES

1. P. J. Phillips, Polymer Crystals, *Rep. Prog. Phys. 53*:549 (1990).
2. A. Lustiger and N. Ishikawa, *J. Poly. Sci., Part B: Polym. Phys. 29*:1047 (1991).
3. S. Bandopadhyay and H. R. Brown, *Polymer 19*:589 (1978).
4. C. J. Singleton, E. Roche, and P. H. Geil, *J. Appl. Polym. Sci. 21*:2319 (1977).
5. P. D. Frayer, P. P. L. Tong, and W. W. Dreher, *Polym. Eng. Sci. 17*:27 (1977).

6. H. R. Brown, *Polymer 19:*1186 (1978).
7. A. Lustiger and R. L. Markham, *Polymer 24:*1647 (1983).
8. N. Brown, X. Lu, Y.-L. Huang, and R. Quian, *Makromol. Chem., Macromol. Symp. 41:*55 (1991).
9. R. N. Haward and D. R. J. Owen, *J. Proc. R. Soc. (Lond.) A352:*505 (1977)
10. E. Ergoz, J. G. Fatou, and L. Mandelkern, *Macromolecules 5:*147 (1972).
11. L. Mandelkern, Golden Jubilee Conference, Polyethylenes 1933–1983: Past, Present and Future, The Plastics and Rubber Institute, London, 1983, pp. D1–D10.
12. R. Popli and L. Mandelkern, *J. Polym. Sci., Part B: Polym. Phys. 25:*441 (1987).
13. J. H. Magill, in *Treatise on Materials Science and Technology* (J. M. Schultz, ed.), Academic Press, New York, 1977, Vol. 10, p. 3.
14. B. Wunderlich, *Macromolecular Physics,* Academic Press, New York, 1973, Vol. 1, pp. 498ff.
15. A. J. Peacock and L. Manderlkern, *J. Polym. Sci., Part B: Polym. Phys. 28:*1917 (1990).
16. M. Furuta and K. Kojima, *J. Macromol. Sci., Phy., B25:*349 (1986).
17. P. I. Vincent, *Polymer 1:*7 (1960).
18. I. Marshall and A. B. Thompson, *Proc. R. Soc. Lond. A221:*541 (1954).
19. F. H. Muller, *Kolloid-Z. 114:*59 (1949); F. H. Muller, *Kolloid-Z, 114:*118 (1949); F. H. Muller, *Kolloid-Z, 126:*65 (1952).
20. D. C. Hookway, *J. Text. Inst. 49:*292 (1958).
21. P. J. Flory and D. Y. Yoon, *Nature 272:*226 (1978).
22. I. M. Ward, *Mechanical Properties of Solid Polymers, 2nd ed.,* John Wiley, New York, 1983, pp. 354ff.
23. I. G. Voit-Martin, R. Alamo, and L. Mandelkern, *J. Polym, Sci., Polym. Phys. Ed. 24:*1283 (1986).
24. Y. I. Matusevich, N. V. Shedova, I. I. Ugolev, and L. P. Krul, *Z. Prik. Khim. 62:*2813 (1989).
25. S. Hashemi and J. G. Williams, *Material und Technik 3:*109 (1986).
26. B. Maxwell, E. J. Dormier, F. P. Smith and P. P. Tong, *Polym. Eng. Sci. 22:*280 (1982).

27. B. Schlund and L. A. Utracki, *Polym. Eng. Sci.* 27:359 (1987).
28. L. A. Utracki, and B. Schlund, *Polym. Eng. Sci.* 27:367 (1987).
29. B. Schlund and L. A. Utracki, *Polym. Eng. Sci.,* 27:380 (1987).
30. S-R. Hu, T. Kyu, and R. S. Stein, *J. Polym. Sci, Part B: Polym. Phys.,* 25:71 (1987).
31. T. Kyu, S-R. Hu, and R. S. Stein, *J. Polym. Sci., Part B: Polym. Phy.* 25:89 (1987).
32. M. Ree, T. Kyu, and R. S. Stein, *J. Polym. Sci., Part B: Polym. Phys.* 25:105 (1987).
33. F. P. La Mantia, A. Valenza, and D. Acierno, *Eur. Polym. J.* 22:647 (1986).
34. J. G. Robinson, M. R. Mackley, and E. Cole, Reinforced polyethylene, European Patent Application 0 351 227 A2, (1990).
35. T. M. Malik, *Polym. Bull.* 26:709 (1991).

8
The Manufacture and Physical Properties of Rubber-Toughened Styrenics

DAVID E. HENTON and ROBERT A.
BUBECK   The Dow Chemical Company,
Midland, Michigan

I. INTRODUCTION

Styrenic polymers are a class of commodity polymers that
have been in commercial production for over 50 years. Polystyrene is the simplest and oldest member of this family, having
been discovered in 1839 [1]. It is generally characterized as
hard, transparent, and brittle. It can be prepared by polymerization of styrene monomer using anionic, cationic, free radical, as well as coordination polymerization processes. One of
the limiting factors in the commercialization of polystyrene
was having an economical route to styrene monomer with
high purity. Although commercial production of styrene mono-

mer was attempted in 1925 by the Naugatuck Chemical Company, real success did not come until The Dow Chemical Company developed the route using catalytic dehydrogenation of ethylbenzene [2]. With the loss of natural rubber sources prior to World War II, the initial incentive to produce high-quality styrene was for use in the Government Rubber Reserve Program to produce synthetic styrene butadiene rubber by emulsion polymerization. With the availability of high-purity styrene, other uses of the monomer were developed including polystyrene.

Styrene can be copolymerized with a variety of other monomers including acrylonitrile, methyl methacrylate, and maleic anhydride. These comonomers modify some of the properties of the polystyrene homopolymer, but do not significantly improve its impact resistance. These materials are amorphous and they fail by a craze formation mechanism followed by craze breakdown and crack propagation. The fracture surface energy for polystyrene [3] is between 250 and 1000 J/m^2, which, for a standard Izod test specimen, is about 0.04 J (0.028 ft-lb, or 0.2 ft-lb/in.) This is very close to the measured impact value of polystyrene and illustrates the brittle nature of styrenic polymers in general. Toughness in styrenic polymers is measured by the ability of a molded article to perform the task the part was designed for, under the environmental conditions of the application. Usually the impact demands put onto parts made with polystyrene and its copolymers are relatively low. Consequently, the use requirements of a styrenic resin usually depend upon stiffness or moisture resistance. Resistance to fracture may often be a secondary requirement.

One of the factors limiting the growth of polystyrene into new applications was its brittleness. It has low impact strength and fails in a brittle fashion at low strain in a flex or tensile test, as well as in a variety of impact tests. One of the earliest attempts to toughen polystyrene was by Ostromislensky, who received a patent in 1927 for a novel approach for combining rubber and polystyrene [4]. He polymerized an unagitated solution of rubber dissolved in styrene monomer. This led to a more flexible product, but the rubber was crosslinked into a continuous phase that prevented it from being processed as a typical thermoplastic. The morphology resulting from these polymerization conditions is similar to that of a sponge where the rubber phase is the continuous sponge and the polystyrene fills the voids in the form of small occlusions (gel particles) [5]. The rubber gels can be broken up by mechanical milling of the product, thus leading to a tougher a more processable material [6]. Other early approaches to toughening polystyrene were mechanical milling of dry styrene butadiene latex rubber with dry polystyrene [7] as well as latex blending of a rubber latex and emulsion prepared polystyrene followed by coagulation and drying [8]. These physical mixtures had improved appearance and processed easily, but the toughness of the products produced by polymerization of styrene in the presence of the preformed rubber was significantly improved over the mechanical and physical mixtures of polystyrene and rubber. This was thought to be because of chemcial bonding (grafting) of the rubber by the polystyrene, which improved compatibility, enhanced occlusion of the rubber by polystyrene, and thus, increased the effective phase volume of the rubber. However, the products made by this process, even when milled, had rubber gels and surface defects. A significant advancement in the technology to make toughened polystyrene came when Amos [9] discovered that by agitating the polymerizing mass of rubber and styrene monomer, dispersed particles of rubber with diameters ranging from approximately 0.5 to 10 μm were formed in a continuous phase of polystyrene. This prevented the formation of a rubber network and resulted in a processable, impact-modified product. This general approach is now the major commercial route to what is called *High-Impact PolyStyrene* (HIPS) [10]. Because of the improvements in the toughness, flow, and appearance of HIPS, it is now used in many demanding applications. It is injection-molded into toys, television cabinets, computer floppy disks, and a variety of household items such as plastic cutlery and cups. It also can be extruded and vacuum-formed into items such as refrigerator liners and cookie trays.

Another outgrowth of the government synthetic emulsion rubber program was the development of ABS (*acrylonitrile-butadiene-styrene* polymer). In this product a brittle styrene-acrylonitrile copolymer is impact-toughened by incorporating a rubber polymer. The early products were based on blends of a nitrile rubber (a copolymer of butadiene and acrylonitrile) and a styrene-acrylonitrile copolymer. These blends can be prepared mechanically or by mixing synthetic latexes of nitrile rubber and emulsion styrene-acrylonitrile copolymer followed by coagulation and drying [11]. These products were first commercialized in 1952 by the U.S. Rubber Company [12], which was formed as an offshoot of the Rubber Reserve Program of World War II. Improvements in the toughness and other properties of ABS came as a result of grafting of the styrene-acrylonitrile onto a preformed rubber substrate, just as in the development of HIPS. Again, the Rubber Reserve Program to produce GRS (a synthetic styrene-butadiene rubber) rubber played an important role. A GRS emulsion-polymerized rubber was used as a grafting base and styrene and acrylonitrile were polymerized in a second step in the presence of the rubber latex [13]. The use of acrylonitrile as a comonomer in the rubber-toughened styrenic often increases the solvent resistance as well as the toughness of the polymer. Other properties are also improved such as stiffness, hardness, and creep resistance. These improvements have put ABS into the class of engineering thermoplastics, and make it the material of choice for many applications.

ABS can also be produced by bulk or bulk suspension processes. In 1964 Dow introduced products based on these processes [2], although the emulsion process is still the most common and produces the majority of the ABS both in the United States and globally. The bulk process has advantages in cost and when low gloss is desired, such as in interior automotive applications. The larger, occluded rubber particles produced by the bulk process inherently provide a matte finish on the surface of molded parts. The emulsion process produces a product with rubber particles typically smaller than 0.8 μm in diameter and less occluded with styrene-acrylonitrile copolymer. This results in a glossier surface in molded articles and thus makes the emulsion-produced ABS resins the products of choice for telephones, power tool housings, and many household consumer products for which appearance is important.

Rubber modification of styrenic polymers is not without its trade-offs. Incorporation of the low T_g polymer reduces the modulus, the hardness, and the scratch resistance of the plastic, and the cost is increased. The viscosity of the polymer is also affected. Rubber-modified polymers have higher viscosities, making them more difficult to mold. Because of the unsaturated elastomer phase, impact-modified polymers are also more susceptible to oxidation and degradation.

The largest volume impact-modified styrenic is HIPS, of which over 30 billion pounds are sold globally. The next largest volume impact-modified styrenic is ABS, which has sales of about 3 billion pounds annually. Smaller quantities of impact-toughened styrene–maleic anhydride copolymer, impact-modified styrene–methyl methacrylate–maleic anhydride terpolymer, impact-modified styrene–acrylonitrile–n–phenyl maleimide terpolymer, and various alpha-methyl-styrene co- and terpolymers are produced.

II. PRODUCTION OF IMPACT-MODIFIED POLYMERS

There are four commonly practiced processes for the production of impact-modified styrenics. Since ABS and HIPS are the two most significant of the toughened styrenics, only processes used to produce them will be discussed, although other toughened styrenic copolymers are made by the same or slightly modified processes. These processes are bulk, bulk suspension, emulsion, and blending.

A. Blending

Blending styrenics with impact modifiers is practiced to a minor extent and almost exclusively to make toughened polystyrene for specialty applications. When high gloss or transparency is needed, blends of polystyrene and styrene-butadiene block polymers are used. By the proper selection of the styrene-butadiene block copolymer, small rubber domains are obtained in the blend that do not scatter light and result in transparency [14]. These types of resins are used for packaging, drinking cups, and lids. Arends [15] found that blends of triblock elastomers and polystyrene result in a toughened resin with a laminar morphology resulting in very high gloss sheet. The composition of the triblock and the viscosity ratio of the block polymer and polystyrene are critical to obtain the proper morphology. High molecular S-B-S triblock elastomers result in particulate morphology with domain sizes of 1 to 3 μm [16], whereas lower molecular weight triblock polymers result in a more laminar morphology. The tensile and flex properties are lowered in the laminar blends because of the co-continuity of the rubbery phase, and flow is increased. Higher molecular weight triblock elastomer blends with polystyrene generally result in greater toughness. Styrene-butadiene block polymers have also been blended with HIPS to improve the flow and gloss of the resin [17]. The area of block copolymer blends with other thermoplastics has been reviewed by Kraus [18].

B. Bulk and Bulk-Suspension Process

The bulk and bulk-suspension process have been reviewed previously with both process diagrams and important operating conditions [19]. HIPS is produced almost exclusively by the bulk process, with some companies combining the bulk process with suspension finishing. HIPS is not produced by the emulsion process because of the difficulty in the emulsion process of producing the large (1 to 10 μm) rubber particles needed to toughen polystyrene. The emulsion process is suited to produce rubber particles in the range of 0.05 to 1.0 μm, making it the preferred process for ABS, which requires a smaller particle for toughness. The bulk or bulk-suspension process is also used for the production of ABS, especially when low gloss is a desired characteristic. The larger rubber particles produced by the bulk process result in a resin exhibiting lower gloss, which is desirable for applications such as the interior of automobiles.

Although a number of different reactor designs and configurations have been patented as previously reviewed, the general concept of the bulk technology can be illustrated using a generalized process configuration (Fig. 1). In this continuous

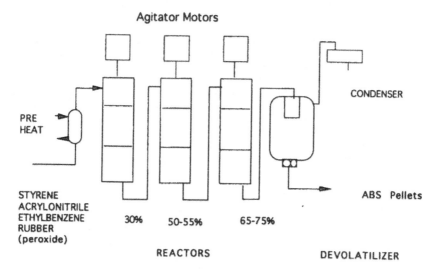

Figure 1 Continuous bulk ABS polymerization process.

process an uncrosslinked, anionically polymerized rubber is ground and dissolved in a mixture of styrene and ethylbenzene along with a phenolic antioxidant, a long-chain alkylmercaptan molecular weight regulator, a peroxide initiator, and optional plasticizers. The rubber solution is continuously pumped from the feed tank into the agitated first-stage reactor where the polymerization reaction starts. The reaction is very exothermic and the heat of the reaction is removed by the cold feeds, through the jacketed wall, internal heat transfer tubes, and, in the case of a boiling reactor, by the heat of vaporization of the monomers and solvents [20]. Phase inversion of the rubber usually occurs in the first stage and the particles are sized from 0.3 to 5 μm depending on many factors. High-rubber-content solutions and high-viscosity rubber generally result in a larger particle size. The ratio between the viscosity of the rubber phase and the viscosity of the styrenic phase is even more important, with a ratio of about 1.0 resulting in the smallest particle size [21]. Echte has quantified the effect of this viscosity ratio by changing the rubber molecular weight at constant polystyrene molecular weight and then by changing the polystyrene molecular weight at constant rubber molecular weight while keeping agitation constant [22]. Agitation speed plays a very key role in controlling the particle size. Dupre [23] has reported the stepwise decrease in the rubber particle size in HIPS from 3.5 μm at 9 rpm to 0.6 μm at 75 rpm. This has a major influence on the toughness and gloss of rubber-modified styrenics.

During the polymerization, grafting occurs on the rubber backbone by either hydrogen abstraction of an allylic hydrogen [24] or addition to a double bond [25]. This grafting reaction is important because the graft copolymer provides the interfacial adhesion between the two normally incompatible polymers and allows the rubber to help initiate matrix ligament bending and crazing in the matrix phase and improve toughness. The graft polymer also acts as an oil-in-oil emulsifer to reduce the interfacial tension in the process and facilitate the sizing of the rubber particles [26]. The choice of initiator for the process balances the need for one that has a half-life in the range of temperatures of the process and one that enhances the grafting process. Oxy radicals are good grafting radicals whereas alkyl radicals are not. Consequently, it has been observed that peroxides such as benzoyl peroxide enhance grafting but azo initiators such as 2,2′-azobisisobutyronitrile, which generate alkyl radicals, do not [27]. Additives in the feed, such as phenolic antioxidants, can negatively affect the grafting process and result in a larger rubber particle size (Fig. 2). The rate of the polymerization as well as the molecular weight of the polymer produced are reduced. The phenolic antioxidant appears to act as a retarder in the process, terminating growing chains as well as reacting with the peroxide to reduce the grafting efficiency, making the process more like that of a thermally initiated polymerization. The larger particles affect gloss and other physical properties. Other factors that reduce grafting have a similar affect on the rubber particle size.

The bulk-suspension process is similar in concept to the bulk process initially and up through phase inversion of the

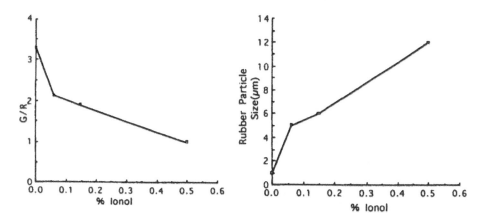

Figure 2 Effect of phenolic antioxidant in the feed mixture on grafting (left) and rubber particle size (right).

rubber and particle sizing. However, instead of operating in a
continuous mode, the bulk-suspension process is operated as a
two-stage batch process. At about 35% solids, the partial polymer is suspended in water using suspension aids such as
polyacrylic acid or polyvinyl alcohol. Polymerization is continued to very high conversion and the polymer stripped to remove residual monomers, dewatered, and pelletized (Fig. 3).
The use of the suspension process eliminates the heat transfer
problems at higher viscosities in the bulk process as well as
the difficulty of pumping and handling the high-viscosity
molten polymer. The rubber particle size and morphology of
products made by the bulk and bulk-suspension processes are

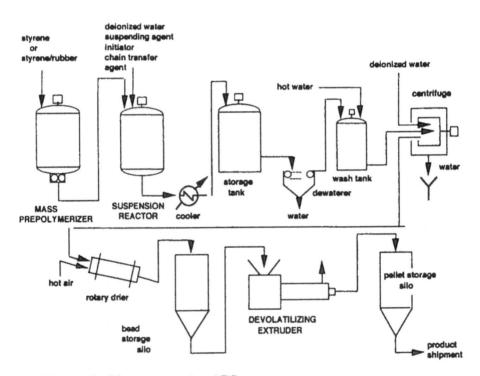

Figure 3   Mass suspension ABS process.

the same because these factors are generally set in the first 35% conversion.

C. Emulsion Polymerization

Emulsion polymerization is the primary process for the production of ABS, with lesser amounts being produced by the bulk process. HIPS is not produced by the emulsion process. In the emulsion process to produce ABS the rubber particles are first produced in a batch process, typically using a water-soluble initiator such as sodium peroxydisulfate, water, emulsifier, and butadiene. These rubber particles are then grafted with styrene-acrylonitrile copolymer (SAN) in a second step, either as a batch or in a continuous emulsion process. In the grafting process the final rubber content could be that of the molding-grade ABS, but for performance and cost reasons a high rubber concentrate is usually produced, which is then let down to a lower rubber level. This is done by latex blending with additional emulsion-produced SAN or melt-blended with SAN produced in a bulk process (Fig. 4).

In the emulsion process the emulsifier is usually a long-chain alkyl sulfonic or carboxylic acid salt such as sodium dodecylbenzene sulfonate or sodium oleate. The monomers are either butadiene or mixtures of butadiene with up to 25% styrene or acrylonitrile. The comonomers increase the rate of polymerization but have a negative effect on the T_g of the rubber, thus affecting the low-temperature toughness of the resin. These factors are balanced depending on the end use of the resin. The rubber particle size necessary for toughness in ABS is larger than 0.3 μm. This size can be obtained directly from the polymerization by using long cycles and a high monomer-to-water ratio and by controlling the amount and timing of the addition of the emulsifier. Small particles are produced by the use of a high water-to-monomer ratio and higher levels of emulsifier. Ott et al [28] have described typical recipes to produce these different-sized particles (Table 1). Emulsion tech-

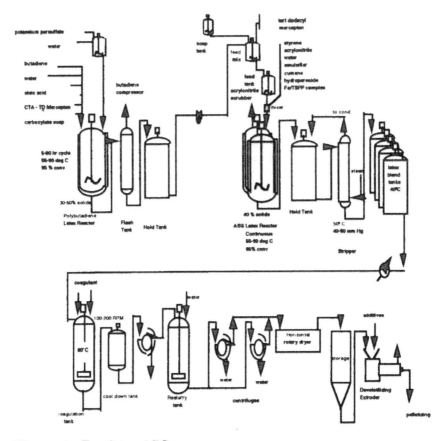

Figure 4   Emulsion ABS process.

nology to produce the base rubber for grafting in the emulsion
ABS process is well known, with refinements by each manufacturer on controlling particle size, crosslinking, grafting,
and isolation.

The rate of polymerization in emulsion is mainly controlled by the particle size of the latex. Since lengthy cycles
are required to produce the large particle rubber, small-
particle rubber latex is often produced in a fast cycle, then

Table 1   Emulsion Rubber Recipes for ABS

	Particle size		
	0.4 μm	0.25 μm	0.1 μm
Deionized water	86	100	200
Initial emulsifier	0.5	0.5	5.0
Total emulsifier	2.0	2.0	5.0
Butadiene	100	100	100
n-Dodecyl mercaptan	0.4	0.4	0.4
Potassium persulfate	0.3–1.0	0.3–1.0	0.3–1.0

partially agglomerated to the larger size by the use of shear [29], neutralization of the emulsifier [30], or addition of hydrophilic copolymer latexes [31]. The crosslink density of the rubber is controlled by the chain transfer agent (mercaptan) concentration, the initiator concentration, as well as the final conversion of the polymerization. Optionally, divinyl comonomers (divinylbenzene) may be added to the monomer feed for additional crosslinking. Higher levels of crosslinking increase the graftability of the rubber and the gloss of the resin, but reduce the high-speed impact resistance. After removal of the unreacted butadiene, the rubber is grafted with SAN to form a core–shell latex. The amount of graft formed in this step is controlled by the ratio of the monomers polymerized to the rubber core as well as the particle size of the rubber. High graft levels can be obtained when a large amount of SAN is polymerized in the presence of a small amount of rubber. Smaller particle have a higher surface-to-volume ratio and thus graft to a greater degree than larger particles. Process conditions as well as initiator and chain transfer levels also affect the graft step. The molecular weight of the graft is mainly a function of the chain transfer concentration (Fig. 5), although the initiator level, polymerization temperature, and other process conditions also have an effect. The amount of

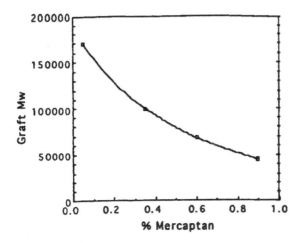

Figure 5   Effect of Mercaptan on graft molecular weight.

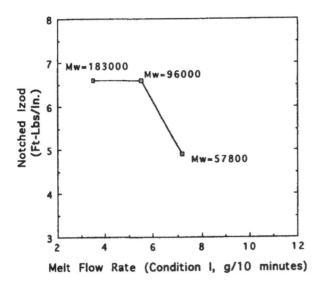

Figure 6   Effect of blending high rubber (50%) ABS resins having varied graft M_w to 16% rubber with an SAN resin having M_w of 93,500.

graft is reduced in proportion to the decrease in molecular weight.

It is well known that the properties of an ABS resin can be varied by blending the ABS to different rubber levels [32]. The balance of melt flow and toughness of an ABS can also be controlled by changing the graft molecular weight (Fig. 6) and the molecular weight of the SAN with which the grafted rubber is blended (Fig. 7).

D.   Stabilization of Rubber-Modified Styrenics

Short-term high-temperature processing and long-term use of rubber-toughened plastics cause oxidative degradation of the rubber phase, resulting in a reduction in impact strength. The

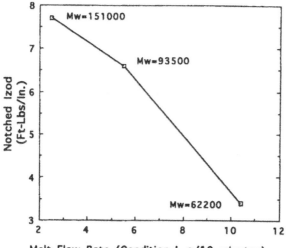

Figure 7   Effect of blending high rubber (50%) ABS with graft M_w of 85400 with SAN resins of different molecular weights to 16% rubber.

degradation occurs through crosslinking, chain scission, and the formation of various oxygenated species on the rubber backbone. Antioxidants are routinely used in these polymers to retard these processes and the technology has been reviewed extensively [33]. Three general classes of antioxidants are commonly used in commercial rubber-toughened styrenics. These include the hindered phenolics, arylphosphites, and thioesters (Fig. 8). Phenolic antioxidants act as radical traps, intercepting the peroxy radical intermediate and generating a stable, nondestructive radical. The effectiveness of these antioxidants is a function of the ease of abstraction of the phenolic hydrogen and is affected by the nature of the substitution on the aromatic ring [34]. Thioesters function as peroxide decomposers and thus decrease the concentration of precursors that lead to the destructive radical intermediates. Phosphites also act as peroxide decomposers and are, in turn, initially oxidized to phosphates. These antioxidants are usually used in combination with each other, resulting in a synergistic effect. One important factor in stabilizing a rubber phase dispersed in a matrix phase is the location of the antioxidant. It must be in the rubber phase in order to prevent the degradation. The distribution of the antioxidant is affected by the polarity of the two phases and that of the antioxidant.

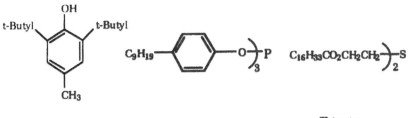

Hindered Phenolic   Arylphosphite   Thioester

Figure 8 Typical classes of antioxidants used in toughened styrenics.

Typically, long chain alky groups are attached to the antioxidant to better match the low solubility parameter of the rubber as well as to reduce its volatility and loss during processing and use.

E.   Other Toughened Styrenic Resins

The use of styrenic resins in outdoor applications puts additional demands, either directly or indirectly, on the long-term durability and properties of the molded article. The ultraviolet light in sunlight is absorbed either directly or indirectly by the benzene ring of the styrene segments inducing a 1S S_O transition [35]. Other adsorbing species such as carbonyl groups are also present in commercial resins, which can contribute to the absorption and degradation processes [36]. This absorbed energy then can lead a variety of radical degradation pathways [37]. Another mechanism has been proposed by Ranby [38] where a charge transfer complex is formed between molecular oxygen and the phenyl group of the styrenic polymer. It was suggested that the complex increases the UV adsorption in the 300 to 400 nm region and provides singlet oxygen, which then plays a role in the radical degradation mechanism. The end result of the UV degradation of styrenics is chain cleavage leading to a reduction in the molecular weight of the polymer, crosslinking of the matrix and yellowing [39]. The crosslinked and low molecular weight polymer on the surface of a weathered rubber-toughened styrenic resin was shown to drastically reduce its impact resistance [40]. In rubber-toughened polymers the unsaturated butadiene rubber at the surface is rapidly degraded leaving holes and flaws that act as stress concentrators upon application of a stress, resulting in a reduction in toughness [41].

Saturated elastomers have been used to toughen styrenic resins when direct exposure to sunlight is involved in the application. The saturated elastomers most frequently used are poly(n-butylacrylate) and ethylene-propylene-diene–modi-

fied (EPDM) elastomers. These elastomers have no double
bonds or allylic hydrogens in the polymer backbone, which are
the sites of ozone, or radical attack in butadiene elastomers. In
the 1960s BASF commercialized a family of acrylonitrile-
styrene-acrylate rubber-toughened (ASA) resins with the
trade name Luran, designed for outdoor use [42]. These products are generally made by the emulsion polymerization route
because of the ease of emulsion polymerization of the acrylate
monomers and the ability to control the molecular variables
in the process. A number of patents claim improvements in
the toughness and properties of ASA resins by controlling the
rubber particle size [43], the rubber particle size distribution
[44], the grafting [45] of the rubber, the crosslinking [46] of
the rubber, as well as the crosslinking of the rigid phase to
encapsulate the rubber particles [47].

AES resins also contain a styrene-acrylonitrile rigid
phase, but use an EPDM elastomer to improve the impact
strength. Uniroyal commercialized an AES resin with the
trade name Rovel in the 1970s for use in applications such as
hot tubs and camper tops. Since the EPDM is produced by a
Ziegler–Natta polymerization process, the AES is produced by
grafting the EPDM in either a synthetic emulsion [48] process
where the rubber is emulsified with monomers, soap, and water, or a variety of solutions [49] or suspension-slurry [50] processes. AES resins generally have higher room temperature
impact strength than ASA resins. They also have a lower
ductile-brittle transition because of the lower T_g ($-55°C$) for
the EPDM elastomer versus $-40°C$ for n-butylacrylate elastomers. ASA resins have better surface gloss because of the
smaller, well-crosslinked acrylate rubber particles. AES resins
maintain their toughness for a longer period of time than ASA
resins in outdoor weathering studies (Fig. 9). Both ASA and
AES resins remain ductile longer than a stabilized ABS resin.
Hybrid blends of AES and ASA have been developed containing large EPDM rubber particles and small acrylate rubber
particles first made and grafted in separate processes and then

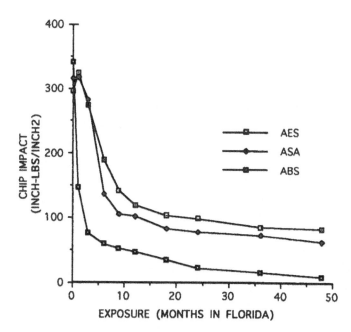

Figure 9   Effects of weathering on the toughness of styrenic resins.

combined. These materials have an improved balance of gloss and toughness [51], as well as good weatherability.

Even by the use of saturated elastomers, good long-term property and appearance retention cannot be obtained by styrenic polymers in outdoor applications without the use of additional stabilizing additives. This is because the elastomer [52], as well as the rigid phase, degrades over time when exposed to sunlight. Therefore, an ultraviolet light absorber such as the o-hydroxybenzophenones or 2-hydroxybenzotriazoles are added to the polymers at about the 0.5% level. Traditional free radical antioxidants such as the hindered phenolics and hindered amines are usually used in conjunction with the light absorbers. Pigments are also important in the long-term property retention of styrenic resins used outdoors [53]. Those that act as UV screeners are very effective in increasing the useful lifetime of the resin, especially titanium dioxide and carbon black. Significant improvements in property retention of styrenic resins are seen when UV screeners and pigments are combined.

Transparent resins can be produced by the use of graft-promoting additives such as allybromide [54], which result in a unique rubber morphology of small rods and capsules. Transparent resins can also be produced by using low refractive index methacrylate comonomers to adjust the rigid-phase refractive index to that of the rubber phase [55]. Commercial resins are based on the refractive index matching approach.

III.   THE PERFORMANCE OF TOUGHENED STYRENICS

Toughened styrenics resins are used in a variety of applications and therefore must be capable of performing under a range of physical and environmental conditions. Many applications require the plastic to maintain its shape and physical properties for many years, often while being subjected to fatigue, chemical exposure, or oxidative attack. In addition to these long-term needs, the toughened styrenic must have sufficient impact resistance to withstand reasonable abuse and not crack or break. A number of standard tests have been developed to test plastics for their long-and short-term performance. Several impact tests are commonly used, including dart-type tests, pendulum impact tests, and variable-rate tensile testing. These common tests are detailed in procedures developed by organizations such as the American Society for Testing and Materials (ASTM).

The properties of commercial styrenic resins range from transparent and brittle (SAN and polystyrene) to tough and solvent resistant (ABS). The addition of rubber to a styrenic resin increases the impact strength and ductility, but reduces

Table 2   Properties of Styrenic Resins

	Tensile (MPa)	E (%)	Flex modulus (MPa)	Izod (J/m)	Vicat (°C)
Polystyrene	40	2	3100	14	106
HIPS	18	40	1930	80	100
SAN	82	3	4100	27	11
ABS	41	20	2300	347	103
SMA	48	2	3612	11	117
Rubber/SMA	32	25	2516	144	115

the stiffness and makes it relatively opaque (Table 2). The incorporation of rubber into a styrenic resin changes the failure mode from a brittle fracture to one with greater ductility at room temperature. With use of the standard impact tests described previously, however, brittle failure can occur in these rubber-toughened systems with increasing the rate of the test (high-speed dart impact) or by lowering the test temperature. This ductile-brittle transition has been shown to be a function of the rubber content in ABS, occurring at lower temperatures with increasing rubber content. The T_g of the rubber also plays a role in the temperature at which the ductile-brittle transition occurs. HIPS containing poly(butadiene-co-styrene) rubber becomes brittle at a higher temperature than HIPS containing polybutadiene, consistent with their relative T_gs (-16°C versus -90°C) [56]. In many applications the temperature at which the ductile-brittle transition occurs is more important than the absolute toughness at room temperature.

In the actual use of toughened styrenics, the component is often exposed to vibrations, repeated low-energy impact, or repeated flexing and bending. Therefore, the fatigue behavior of these materials is important. Fatigue behavior has been studied in great detail, and Sauer and Chen [57] have reviewed this area. More recently, Faitrouni [58] have further investigated some of the effects of molecular weight, rubber content, and rubber particle size on the fatigue properties of ABS. Rubber improves the fatigue resistance of styrenics, as does higher molecular weight rigid phase and adding acrylonitrile to the rigid phase (ABS is more fatigue resistant than HIPS). Amorphous polymers of the styrenic type are not as fatigue resistant as some of the semicrystalline polymers such as the nylons, acetals, and polyolefins.

The addition of a polar comonomer such as acrylonitrile to polystyrene improves resistance of polystyrene to attack by nonpolar solvents and oils. The addition of rubber to styrene-containing polymers also affects the plastic's resistance to aggressive agents. HIPS is more resistant to solvent-induced stress cracking than polystyrene in short duration exposures to n-heptane, although longer exposures show no differences [59]. The effect of rubber morphology is also important in the case of HIPS [60,62], where dense rubber particles or lightly grafted emulsion rubber [61] significantly increase the environmental stress crack resistance (ESCR) of HIPS to food oils. Large rubber particles (i.e., 8 μm) also have been found to increase the resistance of HIPS to failure in the presence of a mixture of oleic acid and cottonseed oil [62]. Other dispersed-phase characteristics such as a high rubber gel fraction [63] have been shown to improve the resistance of HIPS to aggressive agents in applications having low engineering stress. These factors are very important where HIPS is used in applications such as refrigerator liners.

In other applications the ability of a part to support a load without significantly changing dimensions is critical. This load may be simply the weight of the part, or it may be due to the design of the structure. Creep is a long-term phenomenon that occurs in polystyrene via a crazing mechanism [64]. Because rubber in HIPS and ABS greatly enhances the crazing and microshear deformation mechanisms in the rigid phase, the rubber-toughened polymers have poorer creep resistance than their nontoughened counterparts.

IV. MODES OF DEFORMATION FOR TOUGHENING

A. Impact Toughening: A Historical Perspective

Many early theories were proposed to explain the improvement in toughness seen in rubber-modified styrenics. A correlation was seen between the impact strength of a rubber-modified styrenic resin and the peak of the energy loss peak due to the rubber in the −50 to −100°C region as seen in the dynamic mechanical loss spectrum [65]. Because the rubber phase exists above its T_g during normal testing conditions, it was originally believed that the energy of impact was dissipated by rubber phase transitions within the discrete rubber particles, similar to the mechanical energy loss peaks detected by techniques such as the torsional pendulum and the mechanical spectrometer [66]. However, it was calculated that only 10% of the energy of impact could be accounted for by rubber phase transitions [67]. Bucknall [68] was able to show that the whitening that occurred in rubber-modified polystyrene when stressed was due to crazing in the matrix. Microscopy was used to a great extent to probe the nature of the stress whitening seen in rubber-modified styrenics and to try to correlate the relationship between the rubber phase, the matrix phase, and the deformation mechanisms. Phase contrast microscopy was used to observe the crazes generated in HIPS during a tensile loading, which were seen to form around the rubber particles [69]. Osmium tetroxide (OsO_4) staining and electron microscopy was first reported for polymeric systems in 1964 [70]. This technique was applied to ABS [71] and HIPS [72] and was extremely useful in revealing the crazes formed during impact testing of these materials [73]. Using microscopy to observe crazed samples, Matsuo [74] proposed that the differences in thermal shrinkage between the rubber phase and the styrenic phase resulted in a triaxial stress around the rubber particles upon cooling of the plastic melt during a molding operation. This causes the rubber particles to become stress concentrators and then to initiate crazes when an external stress is applied to the plastic. This view is generally accepted today, but disagreement has existed on the sequence of events that leads to the craze formation in the matrix and how crazes are arrested. Kambour and Keskkula [75] proposed that the rubber particles play a dual role, both initiating and terminating crazes. Crazes and cracks radiate as planes perpendicular to the applied force making it difficult for individual rubber particles to arrest a growing craze or crack [76]. It has also been proposed that rubber particles bridge the two sides of a growing crack and provide additional strength and stability to the system similar to that of toughened epoxies [77].

Most recent work on the micromechanism of impact toughening of styrenic polymers has focused on the mechanism of craze initiation—specifically, the order of the events of rubber particle cavitation, crazing within the rubber particle, crazing in the matrix phase, and the relative magnitude of the events in relation to the total dissipated energy. New techniques have been applied to study these phenomona. Laser light scattering has been used to show that rubber particle cavitation precedes crazing in thin ABS samples during tensile testing [78]. Bubeck et al. [79,80] used real-time small-angle x-ray scattering (RTSAXS) to probe the failure mechanisms of a number of rubber-modified styrenics in real time while performing a tensile deformation. This work was the first to show that crazing occurred after the rubber particles started to cavitate, as well as quantifying the relative contributions of energy dissipated by each mechanism. Less than 50% of the plastic strain in ABS and HIPS is due to crazing. An additional mechanism was also proposed as a major contributor to the plastic strain and toughness of styrenics, and that is ligament bending [79]. Ligament bending is the plastic or elastic bending of the polymer matrix between the cavitated rubber particles. Tensile stage microscopy has also been used by Cieslinski [81] to observe the sequence of, first, rubber particle cavitation, followed by crazing in the matrix phase. In this technique thin sections of the rubber-modified styrenic are drawn in a microtensile test device while the deformation is observed by using a transmission electron microscope. By using this technique, Cieslinski also observed rubber fibrilation inside of the occluded rubber particles of HIPS.

B. Stress Fields Associated with Gel–Rubber Particles

A great paradox in the mechanical behavior of brittle polymeric glasses is that an isolated craze or microshear band easily leads to brittle fracture, but the causation of a large number of the same features enhances toughness. The principal purpose of rubber and gel particle modification is to provide for many localized stress concentrations from which to nucleate, propagate, and multiply the possible modes of microdeformation [3,82,86]. Depending on the styrenic matrix, the plastic deformation events that are possible in rubber-modified styrenics are (1) gel and rubber particle cavitation with its associated ligament bending in the matrix, (2) crazing, and (3) microshear yield [3,86]. Postyield extension of rubber and gel particles also occurs [83].

The modeling of gel particles as stress concentraters has been attempted by treating the particle as a concentric shell composite composed of alternating layers of elastomer and matrix polymer. This route has been taken by Matonis [84] and by Manabe and co-workers [85]. A schematic of a concentric shell composite model, as proposed by Matonis, is shown in Fig. 10. From this model, Matonis and Small predicted the radial, hoop, and shear stress distributions that result from a systematic variation of a composite gel particle model upon the application of a stress. The parameters μ_i, ν_i, and R_i are the displacement, Poisson's ratio, and radii, respectively, for each

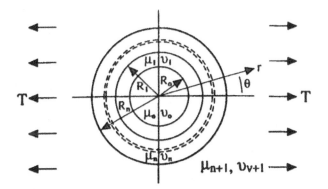

Figure 10 Composite shell model of a gel particle. Depicted is the geometric arrangement and nomenclature for an elastic spherical insert encapsulated by a number (n) of concentric layers of different materials and embedded in an infinite matrix ($i = 1$) under uniaxial tension T. (After Ref. 84.)

shell (i) that composes the composite particle. The analytical result for the radial stress distribution for the model in the surrounding matrix is shown for an isolated particle in Fig. 11. In a typical impact-modified styrenic, the stress fields induced by the particles under thermal and applied stresses interact and enhance the occurrence of the modes of deformation. The multiplication of these deformation mechanisms results in the toughening of the polymer.

C. Deformation Process

The study of plastic deformation of polymers, and materials in general, has often included: (1) examination of the microstructure, (2) deformation of the specimen to the point of fracture, and (3) a postmortem morphological analysis. Although the sequence of deformation events can be determined sometimes by microscopy, microscopy cannot yield the relative contribution of each deformation mode to the total plastic defor-

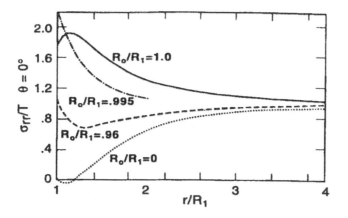

Figure 11 Radical stress distribution associated with a gel particle in a glassy matrix. (After Matonis and Small, Ref. 84.)

mation. Applying the technique of synchrotron x-ray-based real-time small-angle x-ray scattering (RTSAXS), as recently reported [79,80], to the study of relatively high strain rate deformation opens the way to the direct observation of changes in x-ray scattering associated with the modes of deformation.

The principal means applied to the study of deformation of rubber-modified polymers in real time has been quantitative volumetric strain measurements, as practiced by Bucknall and co-workers [86–97] and by Maxwell and Yee [98]. From these measurements the kinetics of craze formation and multiplication during deformation are usually inferred, although a weakness in the technique is its inability to discern crazing from other possible cavitation processes. Indeed, virtually all the dilation in these studies is attributed to crazing alone. Breuer and co-workers [78] performed laser light scattering studies of thin ABS samples during tensile deformation. They showed that rubber particle cavitation associated with localized shear banding preceded crazing in the samples studied.

The purpose of the RTSAXS experiments coupled to video recording of the sample macrodeformation is to ascertain the contributions of the deformation events to the total plastic deformation as a function of time. The scheme for analysis of the data involves the following steps: (1) the determination of total plastic strain ϵ_T from the decrease in x-ray absorption resulting from the sample decreasing in thickness and/or developing internal cavities as a consequence of deformation; (2) the calculation of the plastic strain due to crazing ϵ_{CR} from the analysis of the absolute scattering invariant Q_{Abs} resulting from the scattering from the craze fibrils; and (3) the subtraction of ϵ_{CR} from ϵ_T to obtain the plastic strain due to noncrazing mechanisms (i.e., shear yielding, particle cavitation, etc.) Experimental details are reported in Refs. 79 and 80.).

The use of the optical video information in conjunction with the x-ray analyses permits one to determine the relative contributions of the constant volume mechanisms (elastic, shear, and ligament bending) and of cavitation (exclusive of crazing) to ϵ_{NCR}. The extension ratio in the direction of the sample width Λ_2 is measured from the optical video images of the sample as a function of time. The quantity Λ_3 is proportional to Λ_2 and is a function of sample geometry. From these, the strain associated with the constant-volume mechanisms, labeled here as "shear strain" (ϵ_S) can be subtracted from ϵ_{NCR} to obtain the "remainder strain" (ϵ_R), which is a good measure of the general contribution due to cavitation as one approaches fracture.

The plastic deformation events that were found to be possible in rubber-modified styrenics are plastic strains due to: (1) crazing (ϵ_{CR}), (2) microshear yield (ϵ_S), and (3) gel and rubber particle cavitation and the associated ligament bending (ϵ_R). (*Ligament bending* in this instance is defined as the displacement and deflection of matrix PS or SAN around and in between the cavitating gel particles. Modes 1 and 3 are observed in HIPS and in low rubber mass process ABS resins, and all three modes are observed in the tougher ABS resins.

D.   The Sequence of Events

The results from analysis of RTSAXS impact measurements on 7.5 wt % rubber HIPS at average strain rates of about 7.1/ sec are shown in Fig. 12. The time to fracture is 73 msec and the yield stress is about 3000 psi (20.7 MN/m^2). In the figure the changes in the total plastic strain, the plastic strain due to crazing, the noncraze plastic strain, the scattering center size, and the nominal engineering stress are shown as a function of time. The fact that the plastic strain due to crazing never exceeds 50% of the total plastic strain indicates how important the ligament bending related to the particle cavitation is to the toughness in HIPS. The scattering center size is the average craze fibril diameter, which has been observed to remain at about 13 nm, independent of the decade change in strain rate. Consequently, changes in energy absorption with rate are not influenced by any change in craze microstructure [79,99].

Dynamic characterization of rubber-modified polymers in a transmission electron microscope has recently been performed by Cieslinski [81]. A micrograph by Cieslinski of a film sample of a core–shell rubber particle HIPS during the early stage of plastic deformation is shown in Fig. 13. The micrograph shows cavitation of the rubber shell layer in several particles along with some matrix dilation in an early stage of plastic deformation during which minimal crazing has occurred. These results are consistent with what is observed for the sequence of deformation events derived from RTSAXS analyses for HIPS.

An example of RTSAXS analysis for an ABS at the deformation rate of 0.1/s is shown in Fig. 14. The ABS is a 17 wt % "hybrid" with a 9/1 wt % ratio of emulsion rubber particles to Diene-55 mass process gel particles. The contribution from noncrazing mechanisms greatly dominates that from crazing. No more than a 15% absolute contribution to the total plastic strain from the plastic strain is due to crazing. The average

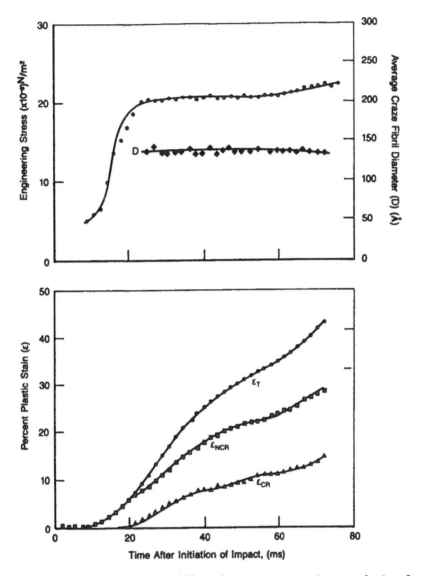

Figure 12 Real-time small-angle x-ray scattering analysis of a typical high-impact polystyrene. The top half of the figure depicts engineering stress and the average craze fibril diameter (*D*) versus time during the tensile deformation of the sample. The bottom half of the figure indicates the total plastic strain and the relative contributions from the noncraze plastic strain and the crazing plastic strain with time. (From Ref. 79.)

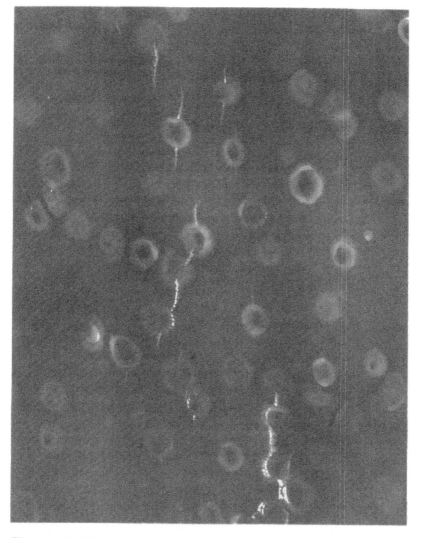

Figure 13   Early stage deformation in a HIPS with core–shell rubber particles (1 cm = 0.25 μm). Transmission electron micrograph courtesy of R. C. Cieslinksi.

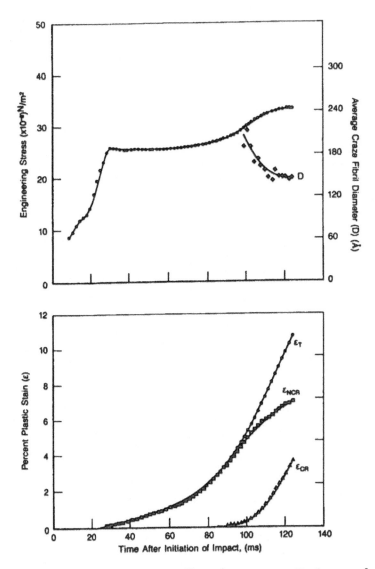

Figure 14 Real-time small-angle x-ray scattering analysis of a typical ABS. The top half of the figure depicts engineering stress and average craze fibril size (*D*) versus time during the tensile deformation of the sample. The bottom half of the figure indicates the total plastic strain and the relative contributions from the noncraze (shear yielding) plastic strain and the crazing plastic strain with time. (From Ref. 79).

craze fibril size in this example is about 18 nm, which is about 50% greater than that for the PS matrix HIPS.

Although the toughest of the studied resins have a *balance* of shear yield, cavitation, and crazing, crazing invariably was the smallest of the contributions to the total plastic strain. Many analyses of dilatation during deformation of rubber-modified thermoplasics attribute most of the contribution to this part of the deformation to crazing [92–94]. In this light a striking result is that the noncrazing strain is, in each instance, greater than that due to crazing. Such results are consistent in kind with the observations made by Breuer et al. [78] for ABS materials. It has been previously observed that shear yielding, presumably associated with particle cavitation, may precede crazing in ABS materials with relatively high impact strength, but not the other way around. As with the HIPS samples, crazing occurs as or after the noncrazing strain commences; however, with ABS, it is believed that the noncrazing strain is most likely due to a combination of microscopic shear deformation (resulting in a change in the lateral sample dimension) and particle cavitation. The contribution due to crazing in the ABS never approaches half of the total plastic strain associated with the deformation.

E.   Craze and Microshear Yield Multiplication

Enhanced toughness via rubber modification is the result of the maximization of contributionss from microdeformation processes. The multiplication of crazing and microshear yield between rubber particles is attained by first of all having a sufficiently high number of rubber or gel particles (i.e., sufficiently high toughening-phase volume content) so that their individual stress fields overlap [80,89]. In ABS, the initial cavitation of particles readily fosters microshear between those cavitated particles provided that they are in close proximity to each other.

It was proposed first by Bragaw [82] that rubber particles are not just craze nucleators, but can cause craze multiplication by craze branching. Donald and Kramer later observed that a single craze can be bifurcated by a gel particle in HIPS [100]. As a craze interacts with a gel particle, it interacts with its stress field and splits. One of the split portions then "walks" up around the particle and establishes a new craze plane above (or below) the original. An important implication of these observations is that the hypothesis [101] that rubber and gel particles can act as craze terminators is an incorrect one.

A strategy for rubber toughening often includes the use of a multimodal particle size distribution. Donald and Kramer observed in a thin-film sample of a hybrid ABS with a combination of small (0.1 μm diameter) emulsion rubber particles and larger (1.5 μm average diameter) gel particles that the small particles cavitated while the large and only the large particles nucleated crazes [89]. This observation suggests that a bimodal distribution of particles induces both the particle cavitation/microshear yield with the many small particles and the particle cavitation/ligament bending/crazing mechanisms in a *balance* so as to achieve consistent toughening performance. The large particles by virtue of their larger associated stress/deformation fields can also serve to "trigger" cavitation in the smaller particles.

V. STRUCTURE–PROPERTY RELATIONSHIPS

Given the material variables available for both the matrix and the impact modification (soft-component) phase, it is possible to tailor the properties of impact-modified styrenics to suit a wide range of engineering applications. Most of the molecular and morphological parameters that affect the technical properties of HIPS and many ABS plastics are listed in Fig. 15, and are schematically represented in Fig. 16 [102]. Of these, the physical properties of tensile modulus and yield,

PROPERTY RELATIONSHIPS

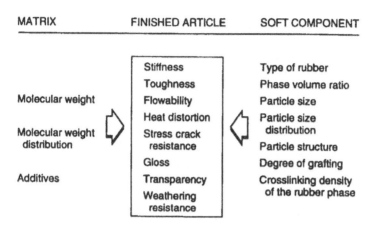

MATRIX	FINISHED ARTICLE	SOFT COMPONENT
	Stiffness	Type of rubber
	Toughness	Phase volume ratio
Molecular weight	Flowability	Particle size
	Heat distortion	Particle size
Molecular weight distribution	Stress crack resistance	distribution
		Particle structure
	Gloss	Degree of grafting
Additives	Transparency	Crosslinking density of the rubber phase
	Weathering resistance	

Figure 15 Molecular and morphological parameters of impact styrenics. (After Ref. 19.)

creep, impact, and environmental stress crack resistance (ESCR) are covered in some detail in the following discussion.

A. Tensile Modulus, Yield, and Creep

The addition of the low-modulus rubber phase to a higher-modulus glassy phase leads directly to decreasing the tensile modulus of the resulting composite. For mass process HIPS containing polybutadiene rubber composite gel particles, Bucknall and co-workers have reported that Young's modulus [93] decreases monotonically from 3.2 GPa to about 1.3 GPa, and that tensile yield [94] decreases from 40 MPa to about 10 MPa over a range of increasing gel particle volume fraction from 0% to 40%. For a given particle type, tensile yield stress (and associated creep resistance) decreases with increasing particle size. For example, the tensile yield stress for a 43% AN ABS with a 16 vol % particle phase fraction decreases from about 72 MPa to 60 MPa, respectively, for 0.4 μm and 1.3 μm average particle

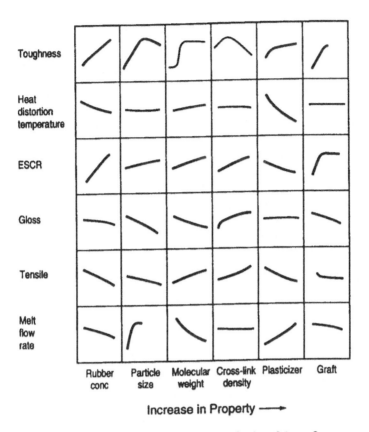

Figure 16   Summary of property relationships of commercial mass process HIPS and ABS. (From Ref. 10, reproduced with permission.)

size [103]. Tensile yield and elongation are also sensitive to deformation rate and temperature, which in turn have a direct bearing on impact behavior.

B. High Rate Impact

Enhancement of impact toughness is the principal reason for impact rubber modification; consequently, impact performance warrants considerable discussion in the context of this review. Two dart impact tests are most frequently used to characterize the toughness of styrenic resins. The drop-weight-type test [104] uses gravity to accelerate a weight that is raised to different heights and dropped onto the plastic part. The weight either has a rounded tip or strikes a dart that punctures the test specimen. The mass of the raised dart and the height can be varied to determine the energy at which 50% failure occurs. High-speed puncture properties are also determined by devices that pneumatically or hydraulically drive a dart through a sample and make measurements using load and displacement sensors [105]. These dart-type tests measure the mutiaxial impact response of the material and can be used to determine the materials sensitivity to rate and temperature. Ductile–brittle transitions are frequently determined in this manner. The data from the driven-dart tests are reported in several different ways. The energy to initiate a crack is often important or, more frequently, the total energy to break the sample is reported.

The two most frequently used pendulum impact tests are the notched Izod [106] and the Charpy. [107] In the notched Izod test, a rectangular molded specimen containing a machined notch is clamped on one end and struck by a pendulum on the other end with the notch facing the striking pendulum. The notch concentrates the stress, minimizes plastic deformation, and directs the fracture to the part of the specimen behind the notch. This test can also be used to determine the sensitivity of a material to notches by varying the radius of the notch [108]. The toughened styrenic resins are not very notch radius sensitive [109] compared to nylons and polycarbonate. The Charpy test is similar in concept to the notched Izod test, using a notched rectangular bar, but the bar is supported on both ends and struck in the center opposite the notch. The energy to break the sample in these tests is reported in units of energy per width of notch (foot-pounds per inch).

The tensile test [110] is also commonly used to measure the toughness of styrenic resins. The absolute strain to failure is used as a measure of the toughness as well as the energy to failure. The rate of the test can be varied to determine the sensitivity of the material to strain rate in a tensile mode. The measured toughness of a sample is a function not only of the type of test used, but also of test temperature, test deformation rate, and sample thickness. As the temperature of the test approaches the glass transition of the elastomer phase, the cavitation process becomes suppressed, thus limiting the mechanical response of the modified styrenic to a greatly reduced amount of crazing. Consequently, the material fractures in a brittle manner with a fracture toughness that is close to that of the unmodified matrix. Such behavior is typical for polybutadiene-modified polystyrene and SAN [3]. The thickness effect on fracture in HIPS with about 7% rubber has been determined by Yap et al. [111]. For thicknesses below about 10 mm, plane stress ductile tearing occurs, whereas above a thickness of 10 mm mixed-mode plane strain-plane stress fracture predominates. Although increasing the deformation rate can be and often has much the same effect as lowering temperature, this need not always be the case for ABS resins. For example, Kobayashi and Broutman [112] have reported nonmonotonically increasing fracture energies for poly(styrene-co-acrylate-co-methyl methacrylate) AMBS copolymers at higher rubber levels.

The mean rubber particle diameter and interparticle ligament spacing can have a strong influence on the toughness as a function of deformation rate. Buckley [80] has reported (Figs. 17 and 18) how toughness under both creep and impact conditions vary with particle parameters, using real-time small-angle x-ray analysis, as described previously. Tensile creep experiments at strain rates of about 0.01/sec and tensile impact experiments at rates of about 50/sec were performed simultaneously with RTSAXS on a series of emulsion rubber particle ABS resins with various average particle sizes and rubber levels. This study provides an example of how rubber

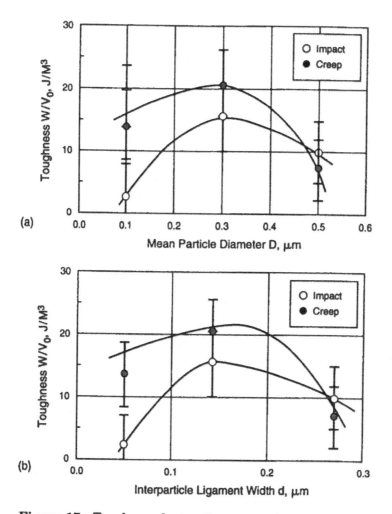

Figure 17   Toughness for tensile creep and tensile impact for ABS as (a) a fraction of rubber particle diameter (D) and (b) interparticle ligament width (d) (After Ref. 80.)

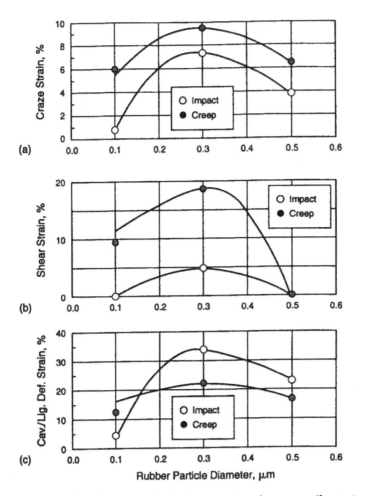

Figure 18 Percent of plastic strains that contribute to the total toughness values reported in Fig. 17 as a function of average rubber particle size and interparticle ligament width. The partial contributions from % plastic strain due to (a) crazing, (b) shear yielding, and (c) cavitation/ligament deformation. (After Ref. 80.)

particle size and spacing interplay with strain rate to result in differing toughnesses, as caused by the contributions from crazing, shear, ligament bending, and cavitation deformation modes. Generally, this series of resins was found to be more tough under the low strain condition (creep) than under the impact condition. Consequently, the testing rate for this series of ABS samples gave the often-observed result of decreasing toughness with increasing strain rate. The observed toughness was maximized at an average particle of size of 0.3 μm, and the plastic strain percentage contributions from crazing, shear yield, and cavitation/ligament bending coincide with this maximized performance. However, the toughness performances were found to be identical for the 0.5-μm particle size and the 0.27-μm interparticle ligament width parameter combination, for which the contribution from the microshear yield mode of deformation is completely suppressed.

C. Solvent-Induced Environmental Crazing

The phenomenon of premature fracture of polymers at stresses considerably below the nominal yield stress while in contact with aggressive environments is a very well established observation, as well as a practical engineering problem [113,114]. This process is a result of enhanced craze growth and breakdown under load and under the influence of an agent that often is only moderately, at most, interactive chemically with the polymer. The degree to which the aggressive agent promotes the formation, drawing, and breakdown of the craze fibrils largely determines the extent to which the critical stress for environment crazing is decreased below the nominal yield stress of the polymer. This fracture process is called environmental stress cracking (ESC). There are two types of ESC, characterized by either a liquid or a gas as an aggressive agent. Liquid environment ESC is given much greater emphasis in the literature and, consequently, is also emphasized here.

An aggressive liquid agent acts upon the polymer at a flaw located at or near the polymer surface and causes localized plasticization and suppression of the glass transition T_g [115] under an applied stress. The critical stress σ_c, at which this process can lead to the onset of craze and crack formation, is considerably less than the yield stress and is a function of the difference between the solubility parameter of the polymer δ_p and of the "solvent" δ_s. The surface tension can also affect values of critical stress [116]. The surface tension of polystyrene against air is 33 to 36 dyn/cm. The effects of surface tension, as well as of molecular size and shape, of the solvent must be considered in addition to δ_s in the assessment of the crazing and cracking tendency of a given aggressive liquid.

Once the deformation zone at the forming environmental craze tip is plasticized and the craze begins to advance, craze growth kinetics is governed by liquid transport through the craze fibril structure where capillary pressure is the driving force [117]. An important implication of these results is that the tips in advancing liquid environment crazes in rubber-modified styrenics are dry, [62] thus indicating that the influence of the solvent on the craze fibrillar microstructure behind the craze tip is very significant in causing premature fracture. The fibril volume fraction v_f in a solvent-grown craze, depending on the environment, can be considerably less than that for an air craze ($v_f = 0.25$) [115]. For example, the T_g and v_f values for crazes grown in ethanol and n-butanol are 85°C and 0.18, and 71°C and 0.09, respectively [109]. The extension of what can be a very limited amount of plasticized fibril matter leads to the premature breakdown and fracture of the solvent-grown craze.

Polymer variables that can affect ESC in styrenics are molecular weight [113,114,118], molecular weight distribution [115], and acrylonitrile content in a styrene-acrylonitrile copolymer (SAN). In many environments SAN is more resistant to ESC than PS; however, simple solubility-parameter concepts are insufficient to fully define environmental behavior because SAN is a polar polymer. Thus, SAN with a $\delta_p = 9.6$ shows a critical stress in n-heptane of 11 MPa. (1570 psi), which is far greater than one would anticipate from solubility parameters alone. This arises because in a completely nonpolar solvent like n-heptane, SAN acts as if its δ_p were greater than 9.6. The excellent resistance of SAN to oils and gasoline accounts for much of its commercial use. In sharp contrast to polystyrene, however, SAN is strongly attacked by methanol. SAN shows good to excellent ESC resistance to n-heptane, cyclohexanol, n-butanol, 2-propanol, and ethanol. It shows poor ESC resistance to decalin, ethyl acetate, amyl acetate, and methanol.

Whereas uniaxial orientation enhances craze and ESC propagation in the direction parallel to the orientation, biaxial orientation suppresses ESC. A principal means of improving the ESC resistance of styrenic polymers is by rubber modification. The polymer parameters used to optimize ESC resistance are, again, molecular weight, rubber particle size and morphology, rubber content, and mineral oil (plasticizer) content [62,118]. If a HIPS sample is placed under constant load in contact with an aggressive liquid such as 50% cottonseed oil–50% oleic acid (to simulate food oil), the times to fracture divide into three zones of behavior, as shown in Fig. 19. The zone of high stress is mainly associated with "dry" craze yield of the sample. For the zone of intermediate stress, time to fracture is generally stress independent because hydrodynamic transport of the aggressive liquid (with capillary pressure as the driving force) through a craze fibrillar structure at the crack tip strongly controls the crack growth rate. The growth rate has a $t^{1/2}$ dependence, [62,117], as shown in Fig. 19. The low stress zone reverts to increased stress dependence because a significant portion of the fracture time is accounted for by an initiation time t_i for crazing as the critical

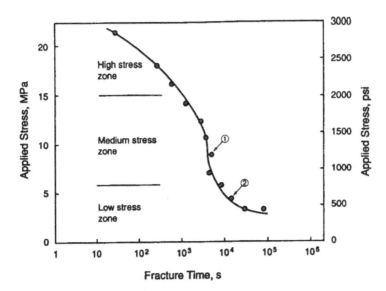

Figure 19 Applied stress versus fracture time for a HIPS sample in contact with 50% oleic acid–50% cottonseed oil showing three zones. Points 1 and 2 refer to Fig. 17. (From Ref. 62, courtesy of *Polymer Engineering and Science.*)

stress for ESC is approached, as illustrated in Fig. 20. The t_i is a result of the necessity to form sufficient dry crazes to permit aggressive liquid transport.

Resistance to ESC is enhanced in HIPS by varying the polymer parameters that govern t_i and govern craze microstructure and stability. The t_i is influenced by the compressive prestress in the polystyrene matrix imposed by the rubber particles [62]. Rubber particles in HIPS are usually "gel particles" composed of the rubber and occluded polystyrene polymer. The thermal prestress arises from the differences in thermal expansion coefficients between the occlusion-filled rubber particles and the relatively unmodified matrix [62,114]. The net compressive thermal stress on the matrix is increased by increasing the volume fraction of the rubber f_r in the polymer, and for a

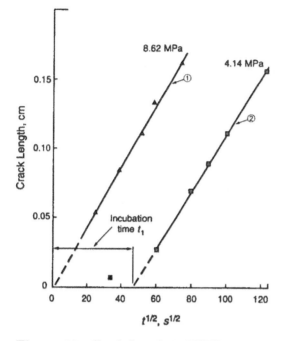

Figure 20 Crack length in HIPS versus square root of time ($t^{1/2}$) at two load stresses, 8.62 MPa (1259 psi) (medium stress zone) and 4.14 MPa (600 psi) (low stress zone) from points 1 and 2, respectively, in Fig. 18 (From Ref. 62, courtesy of *Polymer Engineering and Science*.)

given f_r, decreasing the gel fraction f_g, which is the volume fraction of the rubber plus occlusions. Minimizing f_g maximizes the difference between the thermal expansion coefficients of the matrix and the gel particle as a composite microstructural entity. Increasing f_r and decreasing f_g can raise the critical stress for ESC by several MPa depending on the aggressive environment.

Craze microstructure and stability are controlled primarily by molecular weight, rubber particle size, f_r, and mineral oil content. In commercial practice it is assumed that the rubber or gel particles have "sufficient" crosslinking in the rubber to rubber to endow the particles with the required mechanical tenacity and "sufficient" chemical graft to effectively couple the particles to the matrix. Increasing molecular weight and increasing gel particle size (in the low to medium stress zones) usually improve ESCR. A larger particle nucleates more crazes within its deformation zone, and therefore there are more crazes to accommodate a given strain [114,119]. A large particle is also a better craze stabilizer under load for low strain rate conditions (such as ESC) [120]; thus a large particle helps to retard craze breakdown. It should be noted, however, that proposals of rubber particles being "craze stoppers" during craze/crack propagation are incorrect [119]. Increasing mineral oil content in HIPS decreases ESC resistance because the mineral oil adds to the plasticization of the craze fibrils and because, as evidence suggests, the v_f of the crazes is decreased [114,115], thus facilitating hydrodynamic flow of the aggressive liquid through the crazes. The relative effectiveness of the polymer variables in increasing ESC resistance will vary depending on the aggressive environment. The ESC of HIPS has been studied for various alcohols and alkanes [121,122] and for cottonseed–oleic acid. [62] Similar measurements have also been made for ABS acrylonitrile-butadiene-styrene copolymer (rubber-modified SAN) [123].

D. Fatigue

Fatigue has been defined as the "loss of strength or other important property as a result of stressing over a period of time" [124]. Engineering thermoplastics, as well as materials in general, can undergo brittle fracture after a period of time of experiencing cyclic stresses whose peak values are far below the nominal yield strength. The critical stress below which fatigue failure is not observed to occur is called the endurance limit. Primarily, fatigue failure in styrenic engineering thermoplastics is the result of the formation of crazes and their breakdown under cyclic loading. Higher acrylonitrile content ABS resins often undergo some shear yielding before final fracture by crazing [57,126].

The cyclic stress amplitude has a strong influence on sample lifetime and the mode of specimen failure. A standard way of notating the amplitude is by R-ratio, where R is defined as the minimum displacement (or stress) divided by the maximum displacement (or stress). The maximum is specified as a percentage of strain at yield. Consequently, the severity of the fatigue test increases with decreasing R. The failure lifetimes of polystyrene have been found to fall into two regimes defined by performance below and above $R = 0.6$. [125,126] At R-ratios above 0.6, increases in the R-ratio produce a large increase in fatigue lifetime, whereas at R-ratios below 0.6 similar increases produce much smaller changes in fatigue lifetime and the fatigue lifetimes are much shorter. Real-time small-angle x-ray scattering studies performed during three-point bending fatigue of PS by Brown et al. [126] have shown that in the large R-ratio regime the craze fibrils remain relatively straight during the cycling, but in the small R-ratio regime the load on the craze fibrils becomes compressive and they begin to buckle. Consequently, craze fibril breakdown occurs more rapidly at low R-ratios, and fatigue lifetime is subsequently shortened.

Rubber modification of glassy styrenic matrices usually results in improved fatigue resistance [126]. The generation of crazing and/or microshear results in deformation fields that decrease the effective stress concentration of a growing flaw or crack. Increasing AN level up to about 30% results in increasing fatigue resistance, as does increasing molecular weight. The fatigue resistance of ABS is usually found to be superior to HIPS [57], an example for which has been also been reported by Bucknall and Stevens [127]. This is a result of both the tougher SAN matrix and the greater rubber levels usually found in ABS as compared to HIPS. Tensile yield strengths are usually higher for ABS than for HIPS, and microshear yield in ABS generally tends to precede and dominate crazing, with the consequence that craze formation and breakdown are delayed.

VI. OVERVIEW AND FUTURE DIRECTIONS

This review provides an overview of the methods of manufacture of impact-modified styrenics, and of how variations in their compositions and microstructures determine their physical properties, particularly those pertaining to durability. The choice of a rubber-modified styrenic polymer for a given engineering application usually requires trade-offs between the various desired properties and cost. Ultimately, a balance of properties within the complete context of an application is a prime applications goal. For example, an increased gel particle size in HIPS for a refrigeration liner application may result in better toughness and ESCR, but there may be countervailing reductions in tensile strength (along with creep resistance) and gloss. The origins of such variations in physical properties have been touched upon in this review. Although this review has heavily emphasized physical engineering properties, the appearance of an fabricated article is often a very important, and sometimes, a principal concern. Increasing gel or rubber particle size increases the surface roughness of a part, thus reducing its surface reflectivity of visible light and perceived glossiness. Much future work will probably focus on the maintenance of adequate toughness for a given application while maximizing other properties such as processibility, heat resistance, and appearance.

There has been considerable effort in impact-toughened styrenics over roughly the last 10 years that demonstrates that the energy absorption process during impact consists of a complex combination and sequence of mechanisms. This combination usually includes the process of craze multiplication, but is rarely dominated by it. Aside from crazing, additional mechanisms of energy absorption include ligament bending, particle cavitation, particle extension, and microshear yielding. One implication of these newer findings is that the optimization of impact-toughened styrenics will need to extend beyond the historic routes of the adjustment of molecular weight, and rubber particle size and size distribution. Ligament bending (in HIPS) and microshear yield and ligament bending (in ABS) associated with particle cavitation have been determined to be major sources of impact toughening. Consequently, future toughening strategies will probably be based on changes in particle microstructure and interparticle spacing via increased rubber levels in order to take advantage of the ability of the interactions of particle stress fields to promote these two mechanisms. An important commercial implication of this activity may be enhanced performance at the same or reduced cost of manufacture.

REFERENCES

1. A. J. Warner, R. H. Bounty, and R. F. Boyer, eds. Styrene, Its Polymers, Copolymers, and Derivatives, Reinhold, New York, 1952, p. 3.
2. R. F. Boyer, H. Keskkula, and A. E. Platt, *Styrene Polymers*, John Wiley, New York, 1970, p. 128.
3. C. B. Bucknall, *Toughened Plastics,* Applied Science Publishers, London, 1977, p. 174.
4. I. I. Ostromislensky (to Naugatuck Chemical), U.S. Patent 1,613,673, 1927.
5. G. F. Freeguard, *Polymer 13*:366 (1972).
6. E. D. Morris and G. A. Griess (to The Dow Chemical Company), U.S. Patent 2,606,163, 1952.
7. R. F. Boyer, H. Keskkula, and A. E. Platt, *Styrene Polymers*, John Wiley, New York, 1970, p. 134.
8. R. B. Seymour (to Monsanto), U.S. Patent 2,574,439, 1951.
9. J. L. Amos, J. L. McCurdy, and O. R. McIntryre (to The Dow Chemical Company) U.S. Patent 2,694,692, 1954.

10. R. P. Dion and M. E. Soderquist, *Styrene Polymers,* John Wiley, New York, 1989, p. 88.
11. L. E. Daly, (to U.S. Rubber), U.S. Patent 2,439,202, 1948.
12. C. B. Bucknall, *Toughened Plastics,* Applied Science Publishers, London, 1977, p. 4.
13. C. W. Childers and C. F. Fisk (to U.S. Rubber), U.S. Patent 2,280,773, 1958.
14. R. P. Zelinski and H. L. Hsieh (to Phillips Petroleum), U.S. Patent 3,281,383, 1966, A. G. Kitchen and F. J. Szalla (to Phillips Petroleum), U.S. Patent 3,639,517, 1972.
15. C. B. Arends and M. A. Jones, (to The Dow Chemical Company), U.S. Patent 4,593,063, 1986.
16. S. L. Aggarwal, *Polymer 17:*938 (1976).
17. V. J. O'Grady, (to Monsanto), U.S. Patent 3,907,930, 1975.
18. G. Kraus, *Polymer Blends* Vol 2. Academic Press, New York, 1978, p. 243.
19. A. Echte, *Chemische Techologie,* 4th ed. (K. Harnisch, R. Steinre, and K. Winnacker, eds.), Hanser Publishers, Munich, Vol. 6, p. 373, 1982; A. Echte, Rubber Toughened Styrenic Polymers, *Rubber-Toughened Plastics,* (C. K. Riew, ed.), Advances in Chemistry Series No. 222, American Chemical Society, Washington, DC, 1989, pp. 15–64.
20. W. O. Dalton (Mansanto), U.S. Patent 3,928,495, 1975.
21. H. J. Karam and J. C. Bellinger, *Ind. Eng. Chem. Fund. 7:*576 (1968).
22. A. Echte, *Angew. Makromol. Chem. 58/59:*175 (1977).
23. C. R. Dupre (to Monsanto), U.S. Patent 4,146,589, 1979.
24. J. P. Fischer, *Angew, Makromol. Chem. 33:*35 (1973); A. Brydon, G. M. Burnett, and G. G. Cameron, *J. Polym. Sci., Polym. Chem. Ed. 11:*3255 (1973); A. Brydon, G. M. Burnett, and G. G. Cameron, *J. Polym. Sci, Polym. Chem. Ed. 12:*1011 (1974); V. K. Gupta, G. S. Bhargava, and K. K. Bhattacharyya, *J. Macromol. Sci. Chem. A16:*1107 (1981).
25. R. A. Hayes and S. J. Futamura, *J. Polym. Sci. Chem. Ed. 19:*985 (1981).
26. G. E. Molau, *J. Polym. Sci. Part A1, 12:*67 (1965); G. E. Molau, *J. Polym. Sci. Part A1, 12:*4235 (1965).
27. S. L. Rosen, *J. Appl. Polym. Sci. 17:*1805 (1981).

28. K.-H. Ott, G. Humme, D. Kranz, and H. Rohr (to Bayer), U.S. Patent 4,009,226, 1977.
29. B. D. Jones, Proc. of the 4th Rubber Tech. Conf., London, 1962, pp. 484–497.
30. M. D. Pastor (to Monsanto), U.S. Patent 3,988,010, 1976; M. D. Pastor (to Monsanto), U.S. Patent 4,043,955, 1977.
31. D. E. Henton and T. O'Brien (to The Dow Chemical Company), U.S. Patent 4,419,496, (1983).
32. C. B. Bucknall, *Toughened Plastics*, Applied Science Publishers, London, 1977, p. 298.
33. G. C. Maassen, R. J. Fawcett, and W. R. Connel, *Encyclopedia of Polymer Science and Technology*, Vol. 2. Wiley-Interscience, New York, 1965, p. 171; H. C. Bailey, *Ind. Chem. 38*:215 (1962).
34. G. Scott, *Eur. Polym. J.* (Suppl):189 (1969).
35. G. Loux and G. Weill, *J. Chim. Phys. 65*:484 (1964).
36. P. I. Selivanov, V. L. Maksimov, and E. I. Kirilova, *Vysokomol, Soedin, A11*:482 (1969).
37. B. Ranby and J. F. Rabek, *Photodegradation, Photo-oxidation, and Photostabilization of Polymers*, John Wiley, New York, 1975, pp. 166–190.
38. B. Ranby, *J. Ana. Appl. Pyrolysis 15*:237 (1989); B. Ranby and J. F. Rabek, *J. Polym. Sci., A1, 12*:273 (1974).
39. A. Kaminska, *Polimery, 15*:338 (1970); Also, A. Kaminska, *Polimery, 17*:195 (1972); A. Kaminska, *Polimery, 17*:365 (1972).
40. C. B. Bucknall and D. G. Street, *J. Appl. Poly. Sci. 15*:286 (1968).
41. J. Stabenow and F. Haaf, *Angew. Makromol. Chem., 29/30*:1 (1973).
42. E. Zahn, *Applied Polymer Symposia, 11*:209 (1969).
43. F. Brandstetter, J. Hambrecht, R. Stephan, H. Overhoff, C. Bernhard, J. Swoboda, and A. Echte (to BASF), U.S. Patent 4,442,263, 1984; F. Brandstetter, J. Hambrecht, and R. Stephan (to BASF), U.S. Patent 4,433,102, 1984; J. Swoboda, G. Lindenschmidt, and C. Bernhard (to BASF), U.S. Patent 4,224,419, 1980.
44. D. E. Henton and E. B. Anthony (to The Dow Chemical Company), U.S. Patent 4,753,988, 1988.

45. D. E. Henton (to The Dow Chemical Company), U.S. patent 4,879,348, 1989.

46. D. E. Henton (to The Dow Chemical Company), U.S. Patent 4,801,646, 1989.

47. A. J. Yu and R. E. Gallagher (to Stauffer Chemical), U.S. Patent 3,944,631, 1976.

48. F. J. Limbert and C. F. Paddock (to U.S. Rubber), U.S. patent 3,435,096, 1969.

49. F. X. O'Shea (to Uniroyal), U.S. patent 3,819,765, 1974.

50. W. J. Peascoe, (to Uniroyal), U.S. patent 4,202,948, 1980.

51. D. E. Henton (to The Dow Chemical Company), U.S. Patent 4,766,175, 1988.

52. N. Maecker and D. P. Priddy, *J. Appl. Polym. Sci. 42*:21 (1991).

53. B. Ranby and J. F. Rabek, *Photodegradation, Photo-oxidation, and Photostabilization of Polymers*, John Wiley, New York, 1975, pp. 363–365.

54. C. J. Bredeweg, K. S. Dennis, and C. E. Lyons (to The Dow Chemical Company), U.S. Patent 4,100,227, 1977.

55. J. R. Moran (to Monsanto), U.S. 4,097,555, 1978; J. R. Moran (to Monsanto), U.S. 4,113,798, 1978.

56. C. B. Bucknall and D. G. Street, *SCI Monograph No. 26*, 1967, p. 272.

57. J. A. Sauer and C. C. Chen, *Adv. Polym. Sci. 52/53*:170 (1983).

58. T. A. Faitrouni, Ph.D Thesis, Cranfield Institute of Technology, 1990.

59. R. F. Boyer, and H. Keskkula, *Encyclopedia of Polymer Science and Technology*, Vol. 13. 1970, p. 383.

60. R. A. Bubeck and C. B. Arends (to The Dow Chemical Company), Canadian Patent 1,160,791, 1980; also (to The Dow Chemical Company), U.S. patent 4,521,569, 1985.

61. D. E. Henton (to The Dow Chemical Company), U.S. Patent 4,785,051, 1988.

62. R. A. Bubeck, C. B. Arends, E. L. Hall, and J. B. Vander Sande, *Polym. Eng. Sci., 2*:624 (1981).

63. H. Mittnacht, H. Jenne, K. Bronstert, M. Lieb, H. Adler, and A. Echte (to BASF), U.S. Patent 4,144,204, 1979.

64. C. B. Bucknall, *Toughened Plastics*, Applied Science Publishers, London, 1977, p. 199.

65. L. E. Nielsen, *ASTM Bull. 165*:48 (1950).

66. H. Willersinn, *Makromol. Chem. 101*:297 (1967); Y. Wada and T. Kasahara, *J. Appl. Polym. Sci. 11*:1661 (1967); G. C. Karas and B. Warburton, *Plastic Inst. Trans. J. 30*:198 (1962).
67. S. Newman and S. Strella, *J. Appl. Polym. Sci. 9*:2297 (1965).
68. C. B. Bucknall and R. R. Smith, *Polymer 6*:437 (1965).
69. J. A. Schmitt and H. Kesskula, *J. Appl. Polym. Sci. 3*:132 (1960).
70. E. H. Andrews, *Proc. Ro. Soc. A277*:562 (1964).
71. K. Kato, *J. Electron Microsco. (Tokyo) 14*:220 (1965); also *Polym. Eng. Sci. 7*:38 (1967).
72. R. J. Williams and R. W. A. Hudson, *Polymer 8*:643 (1967).
73. R. F. Boyer, H. Keskkula, and A. E. Platt, *Styrene Polymers*, John Wiley, New York, 1970, pp. 389–391.
74. M. Matsuo, *Polymer 7*:421 (1966).
75. R. P. Kambour and H. Keskkula, *Applied Polymer Symposia, No. 7*, Wiley-Interscience, New York, 1968, p. 215.
76. G. Biglione, E. Baer, S. V. Radcliffe, in *Fracture* P. L. Pratt, ed.), Chapman and Hall, London, 1969, pp. 503–518.
77. S. C. Kunz, P. W. R. Beaumont, and M. F. Ashby, International Conference on Toughened Plastics, Plastics and Rubber Institute, London, 1978, paper 15.
78. H. Breuer, J. Stabenow, and F. Haaf, International Conference on Toughened Plastics, Plastics and Rubber Institute, London, 1978, paper 13.
79. R. A. Bubeck, D. J. Buckley, E. J. Kramer, and H. R. Brown, *J. Mater. Sci. 26*:6249 (1991).
80. D. J. Buckley, Jr., *Toughening Mechanisms in the High Strain Rate Deformation of Rubber-Modified Polymer Glasses*, Ph.D. thesis, Cornell University, Ithaca, NY, 1993.
81. R. C. Cieslinski, *J. Mater. Sci. Lett. 11*:813 (1992).
82. C. G. Bragaw, The Theory of Rubber Toughening in Brittle Polymers, *Multicomponent Polymer Systems*, (Gould, R. F., ed.), *Advances in Chemistry* Series No. 99. American Chemical Society, Washington, DC, 1971, p. 86.
83. A. M. Donald and E. J. Kramer, *J. Mater. Sci. 17*:1765 (1982).
84. V. A. Matonis, *Polym. Eng. Sci. 9*:90 (1969).
85. S. Manabe, R. Murakami, M. Takayanagi, and S. Uemura, *Int. J. Polym. Mater. 1*:47 (1971).
86. C. B. Bucknall, D. Clayton, and W. E. Keast, *J. Mater. Sci. 7*:1443 (1972).

87. C. B. Bucknall, D. Clayton, and W. E. Keast, *J. Mater. Sci.* 8:514 (1973).
88. C. B. Bucknall and I. C. Drinkwater, *J. Mater. Sci.* 8:1800 (1973).
89. C. B. Bucknall and W. W. Stevens, *J. Mater. Sci.* 15:2950 (1980).
90. C. B. Bucknall and C. J. Page, *J. Mater. Sci.* 17:808 (1982).
91. C. B. Bucknall, and I. K. Partridge, and M. V. Ward, *J. Mater. Sci.* 19:2064 (1984).
92. C. B. Bucknall and S. E. Reddock, *J. Mater. Sci.* 20:1434 (1985).
93. C. B. Bucknall, and F. F. P. Cote, and I. K. Partridge, *J. Mater. Sci.* 21:301 (1986).
94. C. B. Bucknall, P. Davies, and I. K. Partridge, *J. Mater. Sci.* 21:307 (1986).
95. C. B. Bucknall, P. Davies, and I. K. Partridge, *J. Mater. Sci.* 22:1341 (1987).
96. C. B. Bucknall, *Toughened Plastics*, Applied Science Publishers, London, 1977, Chapters 7 and 8.
97. C. B. Bucknall, C. J. Page, and V. O. Young, in *Toughness and Brittleness of Plastics* (R. D. Deanin and A. M. Crugnola, eds.,), Advances in Chemistry Series No. 154, American Chemical Society, Washington, DC, 1976, Chapter 15.
98. M. A. Maxwell and A. F. Yee, *Polym. Eng. Sci.* 21:205 (1981).
99. R. A. Bubeck, J. A. Blazy, E. J. Kramer, D. J. Buckley, Jr., and H. R. Brown, *Mater. Res. Soc. Symp. Proc.* 79:293 (1987).
100. A. M. Donald and E. J. Kramer, *J. Appl. Polym. Sci.* 27:3729 (1982).
101. C. B. Bucknall and D. Clayton, *J. Mater. Sci.* 7:202 (1972).
102. R. P. Dion and M. E. Soderquist, *Styrene Polymers.* John Wiley, New York, 1989, p. 95.
103. D. Maes, *Morphology, Deformation, and Fracture Behavior of ABS-Polymers*, Ph.D. thesis, Katholieke Universiteit Leuven, 1989.
104. ASTM Standard D 3029-84, American Society for Testing and Materials, Philadelphia (1984).
105. ASTM Standard D 3763-85, American Society for Testing and Materials, Philadelphia (1985).
106. ASTM Standard D 256-84; Method A, American Society for Testing and Materials, Philadelphia (1984).

107. ASTM Standard D 256-84; Method B, American Society for Testing and Materials, Philadelphia (1984).
108. ASTM Standard D 256-84: Method D, American Society for Testing and Materials, Philadelphia (1984).
109. A. C. Morris, *Plast. Polym. 36:*433 (1968).
110. ASTM Standard D 638, American Society for Testing and Materials, Philadelphia (1971).
111. O. F. Yap, Y. W. Mai, and B. Cotterell, *J. Mater. Sci. 18:*657 (1983).
112. T. Kobayashi, and L. J. Broutman, *J. Appl. Polym. Sci. 17:*2053 (1973).
113. R. P. Kambour, *The Encyclopedia of Polymer Engineering and Science,* 2nd ed. Vol. 4, John Wiley, 1989, pp. 299–323.
114. E. J. Kramer, Environmental Cracking of Polymers, *Developments in Polymer Fracture* (E. H. Andrews, ed.), Applied Science Publishers, Barking, U.K., 1979, Chapter 3.
115. M. B. Yaffe and E. J. Kramer, *J. Mater. Sci. 16:*2130 (1981).
116. H. R. Brown and E. J. Kramer, *Polymer, 22:*687 (1981).
117. E. J. Kramer and R. A. Bubeck, *J. Polymer, Sci., Polym. Phys. Ed. 16:*195 (1978).
118. J. F. Rudd, *Polym. Lette. 1:*1 (1963).
119. E. J. Kramer, *Mechanisms of Toughening in Polymer Mixtures, Polymer Compatibility and Incompatibility Principles and Practices* (K. Solc, ed.), Harwood Academic Publishers, New York, 1983, pp. 251–276.
120. E. Paredes, A. Bustamante, and L. Rivas, *Polym. Eng. Sci. 23:*498 (1983).
121. D. McCammond and S. V. Hoa, *Polym. Eng. Sci. 17:*869 (1977).
122. S. V. Hoa, *Polym. Eng. Sci. 20:*1157 (1980).
123. R. P. Kambour and A. F. Yee, *Polym. Eng. Sci. 21:*218 (1981).
124. R. W. Hertzberg and J. A. Manson, *Fatigue of Engineering Plastics,* Academic Press, New York, 1980, Chapter 1.
125. H. R. Brown, E. J. Kramer, and R. A. Bubeck, *J. Mater. Sci. 23:*248 (1988).
126. J. A. Sauer, A. D. McMaster, and D. R. Morrow, *J. Macromol. Sci. Phys. B12:*535 (1976).
127. C. B. Bucknall and W. W. Stevens, *Toughening of Plastics,* Plastics and Rubber Institute, London, 1978, Paper 24.

9
Polyurethane Polymer Toughness

RALPH D. PRIESTER, JR., D. ROGER
MOORE, and ROBERT B. TURNER*
The Dow Chemical Company, Freeport, Texas

I. INTRODUCTION

This chapter focuses on polymer-toughening methodology for
three of the more demanding polyurethane polymer applications, namely, reaction injection molding (RIM), flexible foams,
and dynamic elastomers (e.g., automotive drive belts, tires, and
shoe soles). The discussions given here attempt to provide an
understanding of the issues associated with each application
area, an overview of the mechanistic understanding, and a
synopsis of the key formulation factors, as well as an overview
of recent technology trends. The information provided is not
intended to represent a rigorous material science approach;

---

*Retired

conversely, it should serve as an introductory primer to those who may be unfamiliar with the toughening aspects of polyurethane polymers. Other application areas such as rigid foams or adhesives and sealants could have been included; however, the issues associated with the latter areas are more pertinent to friability due to high crosslink density or the use of additives to control environmental aging (i.e., covalent crosslink density changes caused by oxidative or ultraviolet degradation) and therefore fall outside the scope of this chapter.

II. REACTION INJECTION MOLDING

A. Rim Fracture Toughness

North American Corporate Average Fuel Efficiency (CAFE) regulations imposed by EPA, which require both weight reductions and a high degree of damage resistance to low speed impact, are promoting the use of plastics by the automotive industry for structural and body applications. Reaction injection molded polymers are the materials of choice because they provide competitive cost and balanced performance for a broad range of shapes and styles from simple bumper covers to full front fascia [1–4]. Similarly, the use of structural RIM composites for noncosmetic applications such as bumper beams and instrument panel supports is rapidly increasing. In either case, it is imperative that the composite be able to manage the energy of on-car impact. Recent cost-modeling studies have highlighted the fact that life cycle repair costs are an important component of the cost differential between polymers and steel.

B. Mechanism

Excellent overviews covering the various modes of failure and the basic theoretical design considerations have been presented previously [5,6].

In the truest sense, the development of new polymers to meet the ever-changing needs of automotive exterior applications is done by the edisonian process. Few predictive models of impact failure are available; the application of fracture mechanics has had limited success. For this reason, the challenge, to date, has centered on the development of a consistent and reproducible methodology by which to compare the range of materials currently in use. In this regard two aspects have received much attention: (1) the criteria needed to provide a representative test and (2) the design variables in the formulation of the polymer.

Even the coarsest inspection clearly shows that on-car impact performance is a complicated mix of many variables [5]. First, the inherent structure of a composite material is highly anisotropic. As a result, stresses and strains and molecular energy absorption mechanism responses to impact loading are highly sensitive to location and orientation. Second, the method of structural support can play a role in the energy dissipation. If the composite is flexible, it may deflect and come in contact with a secondary support, which will absorb the bulk of the impact. In this role the composite must simply be able to bear the force as an intermediary between the impactor and the support. Third, the failure limit of the composite must be matched with the energy dissipation mechanism.

The stresses incurred by a fascia being crushed between a bumper beam and a barrier (i.e., the second situation above) have been modeled Figures 1 to 4 provide a view of the stress fields associated with the compression and subsequent distortion of the part. In all cases the reported values should be viewed as relative rather than absolute. These results highlight the importance of the concentrated tensile stress fields incurred at the radial bends of a typical fascia [7].

Choice of the composite must be made in light of the material's response to impact and the specific application [5]. Figure 5 shows three types of composite material impact responses. The first material is brittle. It undergoes a linear

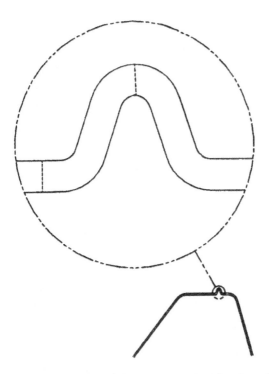

Figure 1　Cross section of RIM fascia with designed radial bend.

elastic deformation and then shatters. The second material is ductile. After an elastic deformation, the material passes through a yield point and undergoes plastic deformation before it reaches the breaking point. The third material, a fiber-reinforced composite, exhibits an abrupt onset of material failure with no plastic deformation. There is a region of fiber breakage and pullout ending with an abrupt break. A homogeneous labeling scheme has been used to characterize the major features of the respective profiles. Referred to as *failure limits*, they can be related directly to the performance criteria required for the specific application.

Thus, the challenge is to increase the level of understanding so that we can expedite the development and characterization of new materials.

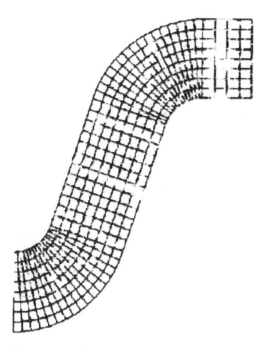

Figure 2 Compression deformation of fascia at radial bend.

C. Applications

Much work has been performed to relate various aspects of RIM chemistry and processing to impact performance. Overviews have been given by Roche, Scammell, Plati, Newman, Martone, and others [8–18]. Important correlations are further discussed below.

D. Hard Segment Content

Variations in the hard segment content are noted to cause significant variations in the range of impact strength observed for a particular part. For polymers having hard segment contents between 40 and 45 wt%, two general trends have been reported: impact strength is found to respond inversely, with hard segment level increases, and the distribution of values narrows as the level increases.

Figure 3   S11 stress contours of fascia radial bend during compression.

E. Fillers

Two standard reinforcing fillers (i.e., milled glass fibers and hammer-milled glass flakes) have historically been used in automotive reinforced RIM (RRIM) applications to provide increased stiffness and improved coefficient of linear thermal expansion (CLTE) and dimensional stability performance [17]. The milled glass fibers are nominally 1.6 mm long with a 16-μm diameter. The standard glass flake is 0.4 mm in diameter and 4 μm thick. The aspect ratio is important because as the

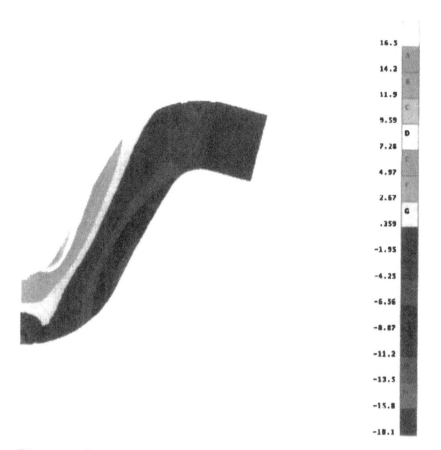

Figure 4 S22 stress contours of fascia radial bend during compression.

reacting polyurethane chemicals are injected into the mold, the filler particles line up with the flow of the liquid mixture. The milled glass fibers point in the direction of the flow; the glass flakes align with the flow, parallel to the surface of the molded part. For this reason, milled glass fibers provide most of their reinforcement in one direction only (i.e., parallel to flow). Conversely, flaked glass produces bidirectionally isotropic properties and is therefore the material of choice for door panels and fenders. More recently, with the advent of high-reactivity,

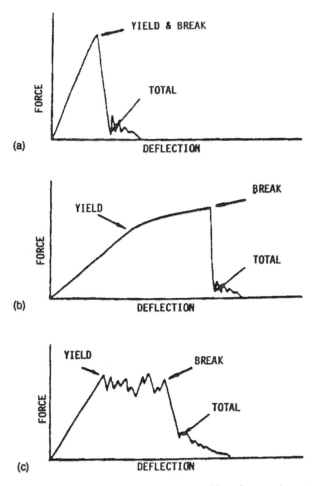

Figure 5 Impact response profiles for various types of composite materials (a) brittle, (b) ductile, and (c) fiber reinforced. (From Ref. 6).

high-performance polyurea systems, mineral fillers such as mica and wollastonite have experienced increased usage (See Tables 1 and 2.).

Unfortunately, in RRIM, increases in modulus often occur at the expense of impact strength. Typical results are presented in Fig. 6 for the unpainted substrate. The addition of

Table 1 Fascia Mechanical Properties

	Tensile strength MPa (psi)	Elongation (%)	flex modulus[a] MPa (kpsi)			Flex modulus ratio (−20°C/70°C)
			At −29°C	At +22°C	At +70°C	
Polyurethane/urea, unfilled	26.7 (3870)	195	849 (123.2)	405 (588.8)	304 (44.1)	2.8
Polyurethane/urea, 15% milled glass	15.2/13.8 (2210/2000)	125/180	618/392 (89.7/56.8)	288/194 (41.8/28.1)	260/137 (37.7/19.9)	2.4/2.9
Polyurethane/urea, 18% wollastonite	13.8/13.3 (1995/1925)	165/180	643/370 (93.3/53.7)	292/182 (42.3/26.4)	241/138 (35.0/20.0)	2.7/2.7
Polyurethane/urea, 15% mica	13.1/12.0 (1900/1745)	205/177	543/413 (78.8/59.9)	245/205 (35.6/29.8)	218/157 (31.6/22.8)	2.5/2.6
High-performance PU/urea, 15% mica	14.9/14.2 (2165/2065)	175/153	463/370 (67.2/53.6)	296/209 (43.0/30.3)	265/179 (38.4/25.9)	1.8/2.1
Polyurea, 15% mica	16.7/15.2 (2425/2210)	155/135	867/743 (125.8/107.8)	407/363 (59.1/52.7)	355/285 (51.5/41.4)	2.4/2.6
Thermoplastic copolyester, unfilled	23.7 (3440)	455	875 (126.9)	354 (51.4)	171 (24.8)	5.1

[a]Parallel/perpendicular.

301

Table 2 Coefficient of Linear Thermal Expansion of Fascia
(CLTE × 10^{-6}/°C)

	−40–22°C[a]	22°–66°C	66°–121°C
Polyurethane/urea, unfilled	148	163	170
Polyurethane/urea, 15% milled glass	49/127	50/139	51/145
Polyurethane/urea, 18% wollastonite	57/134	64/150	68/158
Polyurethane/urea, 15% mica	75/76	85/95	90/105
High-performance PU/urea, 15% mica	75/90	86/102	92/108
Polyurea, 15% mica	72/85	81/96	86/101
Thermoplastic copolyster, unfilled	93/104	131/143	152/163

[a]Parallel/perpendicular.

filler reduces the impact force at failure to about 50% that of the unfilled polymer. Interestingly, the impact force at failure for the painted RRIM composite has been noted to increase over the unpainted substrate. (See Fig. 7.) This improvement has been attributed to increased phase separation as the polymers are subjected to paint oven baking.

F. BINDING AGENTS

Glass fibers come with a multitude of coatings, sizing agents, and binders. These additives are reported to dramatically affect the adhesion between the glass and the RIM resin. Historically, these filler surface modifications have been used to improve adhesion and water resistance in plastics. Surface modifications may also reduce the damage done by mechanical stress in generating surface flaws, even out stress concentrations through some relaxation mechanism, and alter the mode of fracture to failure. An excellent overview of surface modification approaches and their effect on the various modes of fracture failure has recently been presented by Plueddemann [18]. The chemical structure of several of the more commonly used silicone-based coupling agents is given in Table 3. Figure 8 shows electron microscopy photographs

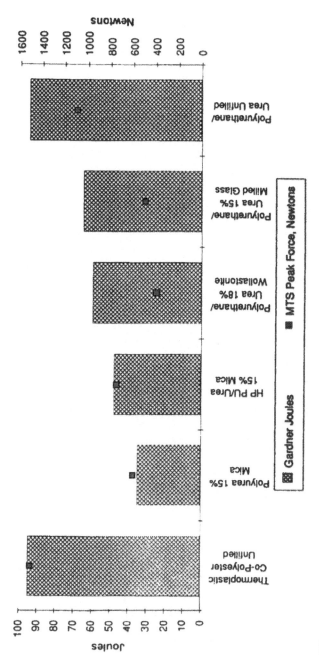

Figure 6 Impact of unpainted fascia at −29°C

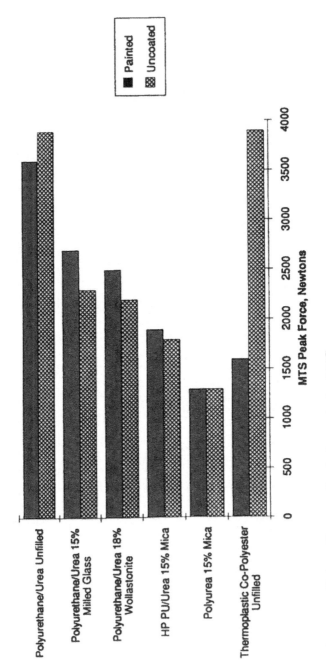

Figure 7 Paint effects on fascia impact at −29°C

Table 3 Representative Commercial Coupling Agents

Organofunctional group	Chemical structure
Vinyl	$CH_2=CHSi(OCH_3)_3$
Chloropropyl	$CICH_2CH_2CH_2Si(OCH_3)_3$
Epoxy	$\overset{O}{\overset{\diagup\diagdown}{CH_2CHCH_2OCH_2CH_2CH_{2Si}(OCH_3)_3}}$
Methacrylate	$\overset{CH_3}{\overset{\mid}{CH_2=C\text{-}COOCH_2CH_2CH_2Si(OCH_3)_3}}$
Primary amine	$H_2NCH_2CH_2CH_2Si(OC_2H_5)_3$
Diamine	$H_2NCH_2CH_2NHCH_2CH_2CH_2Si(OCH_3)_3$
Mercapto	$HSCH_2CH_2CH_2Si(OCH_3)_3$
Cationic styryl	$CH_2=CHC_6H_4CH_2NHCH_2CH_2NH(CH_2)_3Si(OCH_3)_3HCl$
Cationic methacrylate	$\overset{CH_3}{\overset{\mid}{CH_2=C\text{-}COOCH_2\text{-}CH_2\text{-}}} \quad Cl^-$
$N(Me_2)CH_2CH_2CH_2Si$ $(OCH_3)_3$	

Chrome Complex	
Titanate	$\overset{CH_3}{\overset{\mid}{(CH_2=C\text{-}COO)_3TiOCH(CH_3)_2}}$
Cycloaliphatic epoxide	

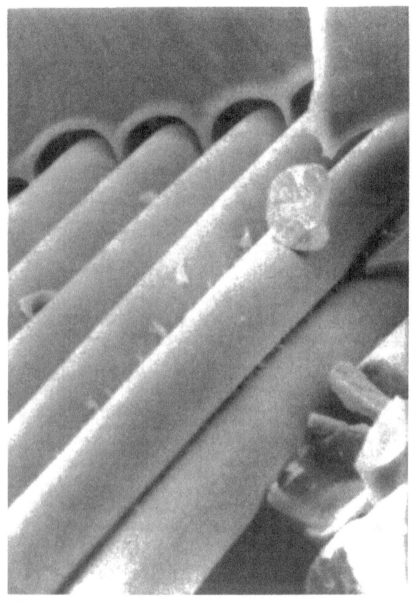

Figure 8 Fracture surfaces of composites after 72 hours in boiling water. (a) Chrome coupling agent. (b) Proprietary coupling agent. (From Ref. 18.)

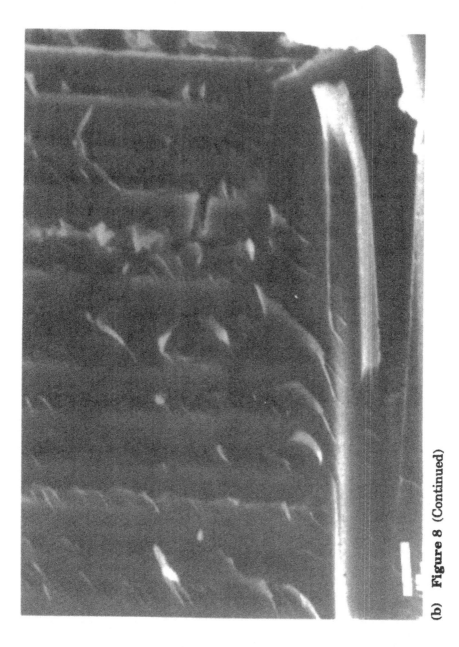

(b)   Figure 8 (Continued)

of the fracture surfaces of two experimental composites. Examination clearly shows the differences in the level of resin–glass adhesion. The results highlight the fact that the glass treatment can have a considerable effect on the final properties of the composite.

G. Structural RIM

Conversely, the use of reinforcing media in structural RIM (SRIM) has been found to improve both modulus and impact performance [19]. Tables 4 and 5 give the mechanical properties of a typical SRIM composite that has been reinforced with three types of filler (i.e., chopped glass, continuous glass mat, and unidirectional glass fabric). Inspection of the results suggests that the unidirectional glass provides the greatest reinforcement coupled with the best impact performance; the performance of the chopped glass and the continuous mat reinforced composites were found to be inferior to that of the unidirectional composite.

Much like the fillers in RIM, use of a coating or sizing

Table 4 Typical Properties of ARSET H12800 Neat and Reinforced with Continuous Glass Mat

Property	% Glass		
	0	22	47
Specific gravity (g/cc)	1.23	1.36	1.55
Tensile strength (psi)	7,000	13,000	28,800
Young's modulus (psi)	320,000	780,000	1,400,000
Elongation (%)	12.0	2.5	2.4
Flex strength (psi)	10,000	20,000	35,100
Flex modulus (psi)	290,000	650,000	1,310,000
Izod impact (ft-lb/in.)	3.5	10.4	10.5
DTUL @ 264 psi (°F)	180	365	410
Hardness, Shore D	80	—	—
Shrinkage (%)	1.35	0.15	0
Heat sag @ 135°C (in)	0.25	0	0

Table 5   Effect of Glass Mat Type on the Properties of ARSET
H12800 Composites

Property	% Glass:	Chopped glass		Unidirec-tional	
		21	48	21	44
Tensile strength, ksi		12	2C7	30	52
Flex strength, ksi		19	35	35	62
Flex modulus, ksi		600	1280	970	2740
Izod impact, unnotched. ft-lb/in.		A8	1A9	35	>890
DTUL @ 264 psi, °F		273	489	336	>500

Table 6   Effect of Glass Treatment on ARSET H12800
Composites

Property	% Glass:	Standard glass mat		Experimental glass mat	
		22	47	23	48
Tensile strength, ksi		13	2B8	13.5	30.6
Young's modulus, ksi		780	1400	845	2000
Flex strength, ksi		20	35.1	24.8	42.2
Flex modulus, ksi		650	1310	780	1580
Izod impact, ft lb/in.		10.4	1C5	C0	21.2
DTUL @ 264 psi, °F		365	410	377	422

agent on the glass mat in SRIM composites is noted to provide
substantial performance improvements. Inspection of Table 6
shows a typical set of physical properties and highlights the
improvements effected by the interfacial adhesion of the glass
to the resin.

H.   Recent Trends

Recent efforts have explored weight reduction by reducing the
cross-sectional area of the part. Key to these studies is the use

of low-monol polyols to provide matrixes having a higher level of covalent crosslinking [20]. Instrumental dart impact testing, conducted on 3.2-mm-thick, painted low-monol polyol samples at −29°C, have dramatically improved impact performance with respect to conventional polyurethane/polyurea polymers having the same level and type of reinforcement. (See Fig. 9.) The mica-filled, low-monol polymers at 4.2 mm thickness have been found to meet the impact requirements for fully painted fascia in original equipment manufacturer barrier impact testing (i.e., 2.2 m/sec; −18°C).

An alternative approach involves the use of highly elastic polymer matrixes that are reinforced with high levels of filler to improve the impact performance [21]. Dart impact methodologies have been used to differentiate the performance of the respective systems; the yield and tear energies were monitored to provide an understanding of impact performance. A typical curve is shown in Fig. 10.

The first part of the curve is a steep linear response representing the elastic deformation of the sample. As the dart moves further into the sample, cracks appear and the load drops sharply. The area under the curve up to this maximum is known as the yield energy. In filled systems the samples yield at the initial fracture point; thus the yield energy represents the energy absorbed without damage. After the yield, cracks are generally noted to grow in a radial direction from the impact point. These events manifest themselves as a hump or plateau after the load drop. Upon complete penetration, sample tearing stops and the load drops to zero. The area under the latter portion of the curve is known as the tear energy.

Figure 11 compares the results obtained from 75,000 psi modulus polymer having 25 wt% filler (HIMOD-LOFIL) with those from a 30,000 psi modulus polymer having 40 wt% filler (LOMOD-HIFIL). Inspection clearly shows that both the yield and tear energies are vastly increased for the latter system, indicating an improved resistance to damage. The fact that

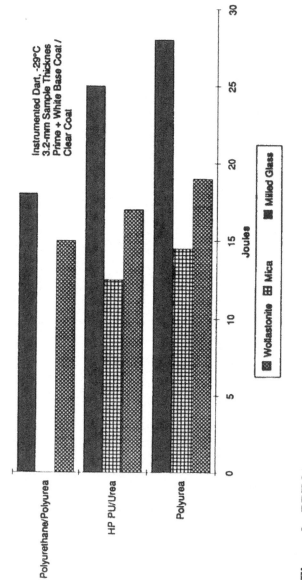

Figure 9 RRIM impact strength (painted).

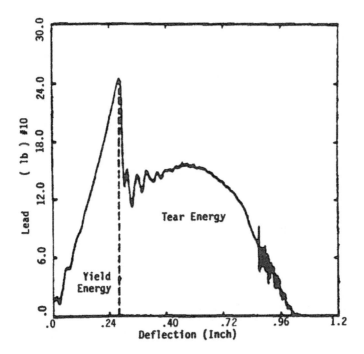

Figure 10 Typical dart impact curve for a high-modulus RIM polymer.

the tear force is higher suggests that once the material is punctured, it is more resistant to tearing. This approach is viewed as a viable route to automotive fascia having the appropriate modulus coupled with vastly improved impact strength.

Automotive engineers are also emphasizing the use of mathematical modeling techniques to identify the complex interactions between the SRIM matrix and the reinforcing mat [22]. As a first approximation in SRIM, density, flexural modulus, flexural strength, tensile strength, and izod impact are determined as a function of wt% glass. A cost-effective comparison of the resultant properties is then made as compared to other composites, aluminum, and steel using existing mechanical property data bases. The models are intended to

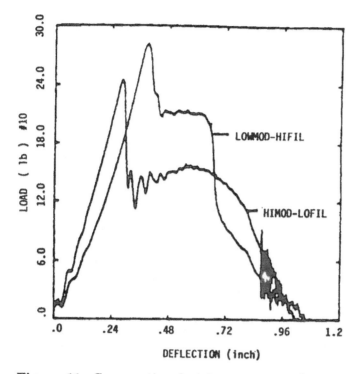

Figure 11 Comparative dart impact curves for a filled polyure-thane RIM polymer versus an unfilled high-modulus polymer.

provide cost-effective first-round screening for the choice of materials and are not intended to replace mechanical testing in situ for the determination of material suitability.

I. Summary

As Corporate Average Fuel Efficiency (CAFE) regulations come into force, the use of composite automotive exterior panels will increase. As this evolution occurs, the importance of improved on-car impact performance from these materials will increase. Today, the polymer development methodology is largely edisonian; however, as the challenge heightens, polyurethane plastics engineers are increasing the use of fracture fundamentals and existing data bases obtained for other materials to design materials having vastly improved performance characteristics.

III.　FLEXIBLE FOAMS

A.　Flexible Foam Fatigue

In typical flexible foam applications (e.g., carpet underlayment, seating, and bedding applications), polymer toughness is synonymous with durability and fatigue performance. This relationship can be rationalized if one remembers that under normal use conditions, the foam is compressed and reexpanded many times. The resultant cyclic stress causes deterioration in foam performance, rendering it unable to provide the degree of cushioning for which it was originally intended. The magnitude of performance loss is the level of durability or the degree of fatigue experienced by the foam. The "real world" quantitation of foam durability is the magnitutude of the indentation load deflection (ILD) or compression load deflection (CLD) [or compression force deflection (CFD)] retention. The CLD measurement is a stress relaxation experiment, in which a foam is compressed to a constant percentage of its original height and allowed to relax over a period of time [23].

1.　MECHANISM

From the preceding discussion, it is apparent that knowledge of the load-bearing behavior of a foamed polymer is needed in order to comprehend fatigue behavior. Four factors must be considered: the covalent network, the hydrogen bonding network, the cell structure, and filler effects. The relative importance of these factors depends on the specific foam. For example, foams containing high loadings of fillers or hard phases might be expected to be more prone to a mechanical degradation or particle cavitation, whereas foams containing large amounts of soft phase would be more prone to failure of the covalent or hydrogen-bonded polyurea network.

Both network factors relate to the breakdown of the polyurea hard-phase hydrogen bonding, which is followed by softening of the hard segments or phase mixing of the hard and soft segments and eventual stress overload of the softer rubber phase [24,25]. In the unstressed state the independent urethane and/or urea chains interact via hydrogen bonding (see Fig. 12). Under stress, the chains shift and new hydrogen

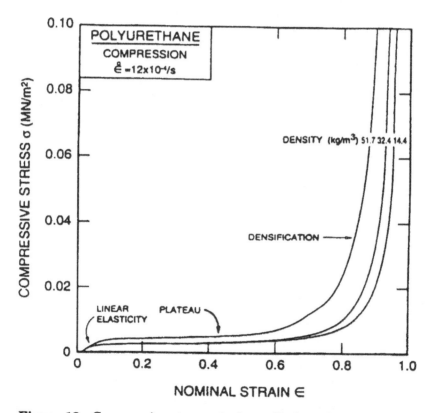

Figure 12 Compression stress–strain profile for a flexible polyurethane foam.

bonds may be formed. Upon removal of the load, the newly formed hydrogen bonds prevent the foam from returning to its original shape. Subsequent load causes increased stress, and mechanical or covalent degradation occurs.

The importance of the hydrogen bond is highlighted by the studies of Kane which show that increases in the 2,6-toluene diisocyanate content (i.e., the isomer providing the highest level of hydrogen bonding) provide improved durability performance [26]. Environmental factors, such as temperature and humidity, can have a pronounced effect on the fatigue performance. Generally, increases in fatigue are noted with increases in either the test temperature or the atmospheric humidity [27,28].

The cell structure effects can be related via correlation of the strut response with the compressive stress–strain curve of a typical foam [29] (see Fig. 13). During the early stages of the curve, the strain is linear and the struts bend in response to the stress. In the intermediate segment the struts are in the

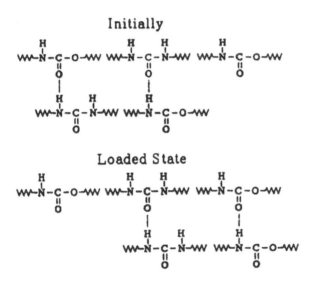

Figure 13   Hydrogen bond rearrangement during fatigue.

process of buckling and the stress response to strain becomes flat. At high levels of compression the struts buckle completely. At that point the stress response increases dramatically and strut breakage is probable. Upon removal of the load, the foam attempts to return to its original dimensions; however, because of the mechanical degradation, it cannot.

The mechanism of strut fracture is a matter of debate. Some studies have indicated that the primary fatigue mechanism is a crazing event in which the buckled struts form small fissures or crazes during the bending process. These fissures continue to grow as the foam is flexed repeatedly until the strut fails. Proposal of this mechanism is based on evidence gathered from acoustic emission, electron microscopy, and penetrant liquid stress depression [30,31].

Various theoretical models have been proposed to explain the stress–strain behavior of foams, including the work of Gent and Thomas [33], Hilyard [34], Rusch [35], Nagy et al. [36], Parma et al [37], and Gibson and Ashby [29]. Gibson and Ashby developed a model for foam mechanical behavior in which the cellular effects could be accounted for by using the square of the density as a scaling factor. Using this relationship, they were able to develop the following relationships to relate the behavior of the foam in the three regimes of the stress–strain curve to the modulus of the polymer:

$$\frac{\text{Foam modulus}}{\text{Polymer modulus}} = \left(\frac{\text{Foam density}}{\text{Polymer density}}\right)^2 \tag{1}$$

$$\frac{\text{Buckling Stress}}{\text{Polymer modulus}} = 0.05\left(\frac{\text{foam density}}{\text{polymer density}}\right)^2 \tag{2}$$

$$\frac{\text{Limiting strain}}{\text{[at densification]}} = 1.4\left(\frac{\text{foam density}}{\text{polymer density}}\right) \tag{3}$$

Using these expressions and further refinements, Gibson and Ashby have been able to predict foam mechanical properties given the polymer density, modulus, and foam density. The impact of foam density in determining the mechanical and durability properties of foam is thus expected to be of paramount importance and will be discussed in a later section.

B. Applications

Much work has been performed to relate various aspects of foam chemistry and processing to fatigue. Overviews have been given by Dwyer, Wolfe et al., Hoffmann et al., and others [38–47]. Table 7 shows a summary of these effects. Important correlations are further discussed below.

1. POLYOL

Several conflicting reports have been published on the effect of polyol molecular weight on durability. Early studies showed that minor changes in the molecular weight or primary hydroxyl content in polyoxypropylene-based triols had little effect on fatigue [43]. Later work showed that in molded-foam systems, increased polyol molecular weight led to poorer foam durability. In carpet underlayment applications, high resiliency (HR) slabstock foams (higher-molecular-weight, ethylene-oxide-capped polyols) have been reported to have better durability than conventional slabstock foams [32,49].

2. ISOCYANATE

Isocyanate type has recently become an area of interest. In the carpet underlayment area, the use of HR slabstock polyols and diphenyl-methylene diisocyanate (MDI) type isocyanates have been reported to give foams that exhibit better durability than conventional slabstock copolymer polyols and toluene diisocyanate (TDI) [32]. The use of TDI with increased 2,6 isomer content has been found to improve durability in static fatigue testing as a result of increased hydrogen bonding for this isomer [26].

The effect of isocyanate index on durability has received a great deal of attention. Most reports agree that indexes significantly greater than 100 result in a deterioration of fatigue performance [38,44]. In carpet underlayment testing it has been reported that increasing foam hardness by the incorporation of copolymer polyol gives significantly better foam durability than increasing hardness by increasing the isocyanate index [42]. This effect has been attributed to less efficient growth of the polymer molecular weight at high index, partially as a result of the excess stoichiometry and partially because of the increase in crosslinking.

3. BLOWING AGENT

The most recent report on the effect of water level was made by Cavender, who studied the durability of HR molded foams of varying foam density [48]. In preparing these foams, the water level was changed, while constant foam hardness was maintained by varying the level of copolymer polyol. This report showed that density and hysteresis were the main factors affecting fatigue, and that lower density gave worse fatigue. This study also showed that hysteresis is related to density, since higher-density foams gave better hysteresis behavior—that is, improved ability to recover from compression. The report indicated, however, that with a constant-density formulation, durability improvements could be achieved by changing the polyols to give a polymer with better hysteresis behavior. Cavender's work also showed the important effect of humidity on the results of durability testing; that is, lower-density foams showed poorer durability under humid conditions than higher-density foams. It should be kept in mind, however, that the composition of the foams was varied during preparation to maintain a constant hardness; thus, the lower-density foams presumably contained higher hard segment levels. This variation may also account for their poorer durability behavior.

Table 7 Chemical and Physical Factors Affecting Foam Durability

Fatigue test method	Variable	Fatigue trend
Symbol meaning	↑ Higher ↓ Lower	↑ Worse fatigue ↓ Better fatigue — No change
Dynamic flexing, roller sheer	Isocyanate level ↑	↑
	Compression set	—
	Density	—
Dynamic fatigue at constant load, 20,000 cycles	Minor changes in MW/1° OH Content	—
	Higher MW polyols ↑	↑
	SAN copolymer ↑	↑
	Isocyanate index ↑	↓
	Water amount ↑	↑ (Slight)
	Inert blowing agent ↑	↑ (Slight)
	Cell structure: fineness, porosity ↑	↓ (Slight)
	Blowing/gelling catalyst ↑	↑
	Density ↓	↓
	Hysteresis ↑	↑
Jounce test	Compression set	Not established
Field trial, human chair occupancy	Resiliency (ball rebound) ↑	↓
	Compression set ↓	↓
	Density ↓	↑
	Tensile, elongation ↑	↓

Roller shear	Density ↑	↓ (Fatigue losses greater around) 1.2 lb/ft³
	Filler ↑	↑
	Firmness (IFD) ↑	↑ (Compared to softer foams at same density)
	HR polyol ↑	—(Relative to conventional polyol)
	Flame retardant	—
	Polyol MW ↓	←
	Functionality ↑	→
Static fatigue, constant deflection	Tin catalyst (DBTDL) ↑	←
	Blowing catalyst: NIAX‰ A-1 ↓	→
	Blowing catalyst: NIAX‰ A-4 ↑	→
	Gel catalyst: DABCO‰ 33 LV ↑	→
	Water level	—(Slight)
	Index	—(Slight)
	Surfactant	—(Slight)

4. SURFACTANTS AND CATALYSTS

Surfactants that promote fine cell structure with a high degree of porosity improve fatigue performance.

The use of blowing and gelling catalyst combinations is standard practice in the manufacture of polyurethane foam. Durability is generally found to be poorer for foams in which the catalyst balance is tilted in favor of one reaction over another.

C. Recent Trends

Recent research into foam durability has attempted to separate fatigue effects that are due to the cellular nature of the foam from those that are due to the polymer. Figure 14 shows

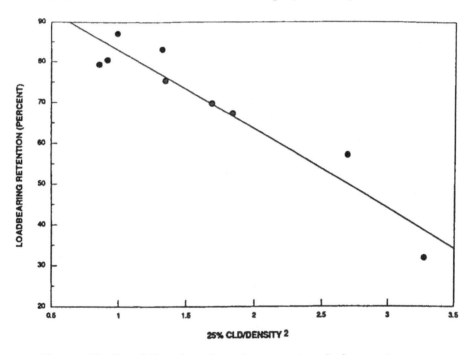

Figure 14 Durability data for prime carpet underlayment.

durability performance data for prime carpet underlayment foams that were submitted to controlled walk testing as part of a cooperative study involving foam chemical suppliers, foam manufacturers, and a major university.

The data are plotted as 25% CLD retention after 160,000 walks versus 25% CLD divided by the square of density. The square density function is derived from the theory of Gibson and Ashby, which shows that the actual stress on a foam is a function of density squared [29]. The linear relationship shows the interaction of hardness and density in determining fatigue. Figure 15 shows data from Ref. 46 plotted in the same manner, indicating that this relationship has validity for data from several sources.

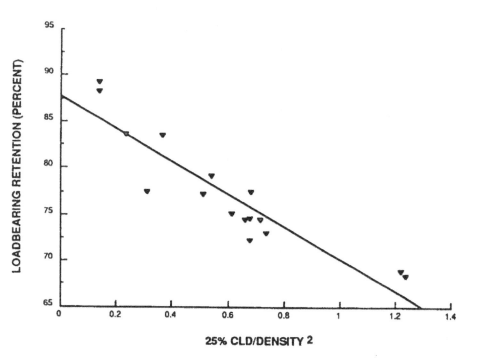

Figure 15 Load-bearing retention data for prime carpet underlayment foams.

The main conclusion to be drawn from this type of analysis is that the most important cellular factor influencing carpet cushion fatigue is density. These data show that foams with a density of less than 2 lb/ft³ would be more likely to suffer catastrophic load-bearing failures after only limited in-service use. In order to decouple the cellular effects from the polymer effects, the data from this study were further analyzed by normalizing the load-bearing retention by dividing by the 25% CLD over density squared factor. This normalization removes the cellular effects from the data and allows a correlation to be made of durability behavior to polymer properties. Figures 16 and 17 show the normalized load-bearing retention plotted as a function of foam hysteresis (via ASTM

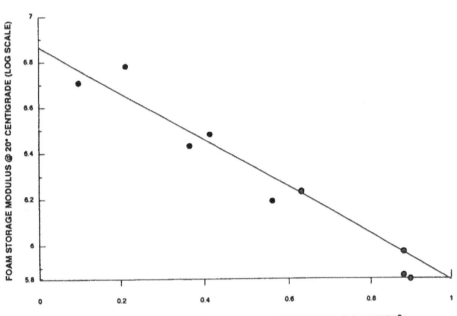

Figure 16 Foam storage modulus related to load-bearing retention in prime carpet underlayment foams.

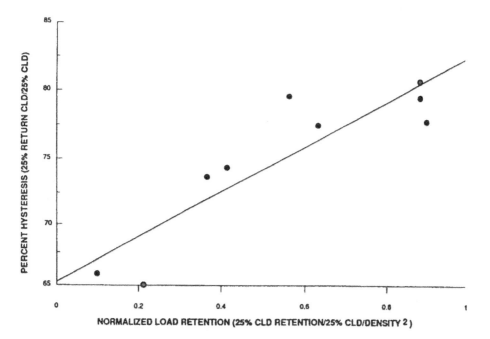

Figure 17 Foam hysteresis related to load-bearing retention in prime carpet underlayment foams.

D 3574) and foam storage modulus determined from dynamic mechanical spectroscopy.

These parameters, which are more directly related to the molecular properties of the polymer in the foam, show some interesting effects. Foams with lower hysteresis (worse ability to recover from compression) have worse durability than high-hysteresis foams, and foams with higher storage modulus (stiffer polymer in the struts) have worse durability than softer polymers. High storage modulus and low hysteresis are characteristics associated with foams that contain high levels of hard segments (either polyureas or styrene-acrylonitrile (SAN) copolymer particles). Lower-density foams which are all water blown will contain higher levels of polyurea than higher-density foams. This gives a negative durability effect both in lowering density and in raising hard-segment content. Adding SAN copolymer polyol to low-density foams will lead to higher hard-segment content and therefore cause the durability to deteriorate.

In the European market, methylene diphenyldiisocyanate (MDI)–based technology is predominant. Because of the many differences incurred with the change in isocyanate, the research and development focus has been to obtain a basic understanding of network formation and to correlate the relative contributions of the physical and covalent bond formation with polymer performance. Key to this effort is the understanding of the importance of polyols having low levels of monol. Multiple polymers, based on systematic variations of water level and/or polyol and isocyanate functionality, have been compression molded (to negate cell structure effects) and examined for their elastic properties. Characterization techniques include static stress–strain measurements using both constant load and constant deformation approaches and traditional creep experiments. In general, the network integrity of the polymers was found to correlate in a semiempirical manner to the Mooney–Rivlin equation:

$$f^* = 2C_1 + \frac{2C_2}{L} \tag{4}$$

where

f^* = reduced force
L = extension ratio
C_1, C_2 = Mooney–Rivlin constants

The interpretation of the constants C_1 and C_2 has absorbed much time; the results are inconclusive. Previous studies have suggested that the ratio $2C_2/2C_1$ is related to the looseness with which the crosslinks are embedded within the structure. For example, trifunctional crosslinks have a larger value than tetrafunctional crosslinks. Conversely, the constant C_2 decreases with an increasing degree of solvent swelling. Many indications suggest that C_2 is more a function of entanglements between chain segments; thus the constant is noted to approach zero for more uniform networks.

In the MDI-based foams the constant C_1 was found to increase with increases in both the covalent and physical bonding. The constant C_2 was found to relate to network homogeneity and decreased with decreasing polyol functionality. Deviations from theoretical behavior were noted at high levels of urea content; weight fraction hard-segment corrections to the Mooney–Rivlin plots appeared to provide improved relationships. Creep experiments support the sensitivity to urea content; samples with high urea content were noted to exhibit very steep modulus decay as a function of time. Creep moduli at low urea content were observed to be 10 to 40 times lower than those of the high urea samples. Differences in behavior versus the extent of covalent crosslinking in the high urea content samples were found to be negligible in comparison to the urea content effect.

D. Summary

In summary, several factors are found to be important in foam durability. The current understanding of foam durability suggests that the effects due to the cellular nature of the foam and the effects due to the polymer must each be accounted for in order to develop a better understanding of their roles, both individually and together, in assessing foam durability. It is only through such studies that a firm theoretical foundation can be built so that further foam durability improvements can be made.

IV. DYNAMIC ELASTOMERS

A. Elastomer Dynamic Fatigue

Many different types of polyurethane-based elastomeric materials are currently available. Typical formulations are based on only three components (i.e., a diisocyanate, a polyol, and a chain extender), yet the number of combinations is unlimited. Their demonstrated ability to provide both a wide range of hardnesses and an excellent overall set of physical properties coupled with chemical, solvent, and abrasion resistance and biocompatibility make these polyurethanes the materials of choice for demanding applications, such as solid wheels and rollers, automotive drive belts, and shoe soles. In these applications the polymer experiences dynamic stress at a variety of frequencies; part of the resultant dynamic energy is dissipated in the form of heat. Because the polyurethane elastomers have a low thermal conductivity, the generated heat causes an increase in the temperature of the material, which in extreme cases causes the elastomer to exceed its thermal limits.

1. MECHANISM

Although a knowledge of the dynamic properties of polyurethane elastomers is critical to their use in dynamic stress applications, little has been published. Recently, two groups have attempted to characterize these materials in terms of high-temperature mechanical properties using existing dynamic mechanical testing protocols [50,51].

As with flexible foams, when a cyclic load is applied to a polyurethane elastomer, not all the energy is recovered when the load is removed. Some of the mechanical energy is converted to heat. This is known as hysteretic energy loss and is represented by the area between the load and relaxation curve (see Fig. 18). In a perfectly elastic polymer, which stores all the input energy, the level of heat generation would approach zero. However, in "real-world" polymers, a viscous or loss response component must be included. This component, which

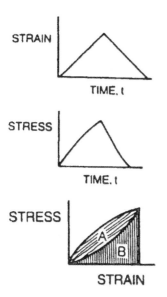

STRAIN

TIME, t

STRESS

TIME, t

STRESS

A

B

STRAIN

Figure 18 Principles of the stress–strain curve.

arises from molecular reorganization or motion, causes heat buildup that leads to covalent degradation of the matrix and loss in performance properties.

Dynamic mechanical data can be used to estimate heat generation in practical applications where a material undergoes cyclic deformation [54,63]. The heat generated per cycle at a specific peak strain is directly proportional to the loss modulus G'' as shown in Eq. (5). $\mathring{a}_0$ is the maximum value of the shear strain during a cycle.

$$H = \pi G'' \mathring{a}_0^2 \tag{5}$$

When a material undergoes cyclic deformation at a specific peak stress loss compliance, J'' is the relevant function and the heat generated per cycle is given in Eq. (6), where Π_0 is the maximum value of shear stress during a cycle.

$$H = \pi J'' \Pi_0^2 \tag{6}$$

B. Applications

Much work has been performed to relate various aspects of polymer chemistry and processing to dynamic fatigue performance. Overviews have been given by Clift, Nielsen, Sung, Kogelnik and others. [50,53–63] Important correlations are discussed below.

1. SOFT SEGMENT POLYOL STRUCTURE AND MOLECULAR WEIGHT

Polymeric diols commonly used in polyurethane elastomers are typically based on polytetramethylene ether glycol (PTMEG), polyoxypropylene glycol (PPG), or polyesters. At 1000 molecular weight, the soft-segment glass transition temperature increases in the order of PTMEG < PPG < esters. This order results from both the inherent glass transition of the material and the increased phase mixing due to increases in the level of soft-segment/hard-segment compatibility. Rubbery moduli and hard-segment melting points decrease in the order of PTMEG > esters > PPG.

Decreases in the molecular weight of the PTMEG chain provide increased levels of phase mixing. Soft-segment crystallinity is reduced and the storage modulii is increased because of the increase in hard segment content. Conversely, increases in the molecular weight of the PPG chain are noted to reduce the level of phasing mixing and provide improved hard-segment order.

Cast elastomers prepared with high molecular weight polyols (molecular weights between 6000 and 10,000) containing low-monol functional content have demonstrated higher hardness values and significantly higher strength properties than elastomers prepared with conventional polyols [63].

2. CHAIN EXTENDER STRUCTURE AND LEVEL AND DEGREE OF POLYMERIZATION

The choice of chain extender has considerable influence on elastomeric properties. The selection of an appropriate curative must incorporate its reactivity, ease of handling, toxicity, and cost. An excellent overview of these factors and the effect of the degree of polymerization on thermal stability and phase separation, for both glycols and amines, has been published by Christenson and co-workers [64].

In amine chain extenders (the more commonly used extenders include 4,4'-methylene-bis-orthochloroaniline (MBOCA), 4,4'-methylene-bis-[3-chloro-2,6-diethylaniline] (MCDEA), di-(methylthio)-toluenediamine (Ethacure* 300), trimethylene-glycol-di-p-aminobenzoate (Polacure† 700), the structural symmetry is noted to have a pronounced effect on both the level of phase separation and the ultimate thermal stability. Hard-segment melting temperatures decrease in order of MCDEA < MBOCA < Polacure < Ethacure 300.

The storage modulus is generally found to be a function of the hard-segment level.

3.   ISOCYANATE ISOMER DISTRIBUTION

Highly symmetrical molecules, such as 2,6-toluene diisocyanate (TDI) or 4,4'-methylene diphenyldiisocyanate (MDI), are noted to provide highly phase-separated elastomers with lower soft-segment glass transition temperatures, a higher storage modulus, and higher hard segment melting points. Lower loss compliance values for this type of structure would predict improved dynamic properties with lower heat generation.

C.   Recent Trends

Recent efforts have targeted three elastomer applications areas—automotive drive belts, tires, and shoe soles. North American research has attempted to develop a better under-

---

*Trademark of the Ethyl Corp.
†Trademark of Air Products.

standing of dynamic fatigue by studying the crack growth characteristics of the polyurethane-polyurea elastomers used for tire and drive belt applications. Representative samples were hand cast into plaques and evaluated using an MTS servo-hydraulic system. The critical load, the load required, to cycle the material over a 2500 to 5000 cycle range, was determined with respect to both part thickness and use temperature. Both critical load and hysteresis varied linearly with part thickness. In general, no differentiation in critical load was observed in ambient temperature testing; significant differences were observed in 100°C testing.

Crack growth rates were measured at both ambient and elevated temperatures. Crack length was monitored as a function of the number of cycles. Data analysis consisted of measuring fracture toughness, K_{1C}, as a function of geometric factor, critical load, and crack length. In general, all materials exhibited the same crack growth rate during ambient testing. Polymers exhibiting stress crystallinity distinguished themselves at elevated temperature by exhibiting low crack growth rates.

The effect of formulation variation, processing, and annealing on polymer durability was evaluated. No improvement in the cycle life was obtained at postcure temperatures up to 120°C. Increases in % NCO (hard-segment content) provided only modest improvements in cycle life.

In the European market, flex fatigue resistance and abrasive wear at the most critical parameters in the qualification of materials for shoe soles. Recent legislation has mandated a shift from chlorofluorocarbon systems to all-water-blown formulations. This change has caused a deterioration in these properties. Material science evaluations clearly show that this effect is due to increases in phase mixing effected by the in situ formation of urea. These detriments are circumvented by incorporation of ethylene oxide into the polyol. This modification effects improved polyol-water compatibility, which restores the blow—gel balance.

V. SUMMARY

Polyurethane elastomers are well known for their exceptional abrasion and tear resistance, high load-bearing capability, and design flexibility. However, the demands of the new in-service applications are rapidly increasing. Applications such as automotive belts, tires, and wheels require that the polymer withstand high-temperature, high-stress conditions for extended periods of time. The key to new polymer development for these applications is an understanding of the relationship between morphology and performance. Dynamic mechanical spectroscopy has proved to be a powerful tool for these applications.

ACKNOWLEDGMENT

The authors wish to acknowledge the graphics assistance provided by Elaine Sonnier, Bob Bowden, and Paul Neill in the preparation of this chapter.

REFERENCES

1. W. F. Gum, W. Riese, and H. Ulrich, *Reaction Polymers*, Hanser Publishers, New York, 1992.
2. W. E. Becker, *Reaction Injection Molding*, Van Nostrand Reinhold, New York, 1979.
3. M. C. Cornell, Whither Bound RIM? *Proceedings of the SPI 33rd Annual Technical/Marketing Conference*, Technomic, Lancaster, PA, 1990, pp. 442–427.
4. G. Slocum, Reaction Injection Molding, *Composite Material Technology: Processes and Properties* P. K. Mallick and S. Newman, eds., Hanser Publishers, New York, 1990, p. 103.
5. A. E. Mascarin, Fender Material Systems: A Lifetime Cost Comparison, SAE International Congress and Exposition, Paper 920373, 1992.

6. S. L. Kessler, G. C. Adams, S. B. Driscoll, and D. R. Ireland, *Instrumented Impact Testing of Plastics and Composite Materials*, American Society for Testing and Materials, Philadelphia, 1986.
7. A. Moet, The Crack Layer Theory: A Unified Approach to Fatigue and Fracture Toughness, presented, in part, at the International Symposium on How to Improve the Toughness of Polymers and Composites, Yamagata, Japan, October 1990.
8. J. L. Roche and S. N. Kakarala, Methodology for Selecting Impact Tests for Composite Materials in Automotive Applications, *Instrumented Impact Testing of Plastics and Composite Materials* (S. L. Kessler, G. C. Adams, S. B. Driscoll, and D. R. Ireland, eds), American Society for Testing and Materials, Philadelphia, 1986.
9. K. L. Scammell, Comparison of Strain Rates of Dart Impacted Plaques and Pendulum Impacted Bumpers, SAE International Congress and Exposition, Paper 870106, 1987.
10. E. Plati and J. G. Williams, *Polym. Eng. Sci. 15*:470 (1975).
11. D. G. Cook, A. Rudin, and A. Plumtree, *Polym. Eng. Sci. 30*:596 (1990).
12. J. G. Williams, *Polym. Eng. Sci. 17*:144 (1977).
13. T. A. Vu-Khanh and F. X. De Charentenay, *Polym. Eng. Sci. 25*:841 (1985).
14. G. C. Adams, R. G. Bender, B. A. Crouch, and J. G. Williams, *Polym. Eng. Sci. 30*:241 (1990).
15. R. W. Hertzberg and J. A. Manson, *Fatigue of Engineering Plastics*, Academic Press, New York, 1980.
16. L. V. Newmann and J. G. Williams, *Polym. Eng. Sci. 20*:572 (1980).
17. R. E. Morgan, J. W. Berg, and G. A. Klumb, Polyurethane/Polyurea Polymers for High Performance Automotive Fascia, *Proceedings of the SPI 34th Annual Technical/Marketing Conference*, Technomic, Lancaster, PA, 1989, pp. 262–268 and references therein.
18. E. P. Plueddemann, *Silane Coupling Agents*, Plenum Press, New York, 1982.
19. U. E. Younes, High Impact Structural RIM Composites. *Proceedings of the SPI 34th Annual Technical/Marketing Conference*, Technomic, Lancaster, PA, 1989, pp. 292–296.

20. B. D. Shrott and K. F. Bennett, Downgauging with High Performance, Filled RIM Polymers—Lowering Fascia Cost and Improved Quality, SAE International Congress and Exposition, Paper 920528, 1992.

21. D. M. Rice, R. A. Grigsby, and C. S. Henkee, Recent Advances in Filled Polyurea RIM, *Proceedings of the SPI 34th Annual Technical/Marketing Conference*, Technomic, Lancaster, PA, 1992, pp. 231–237.

22. G. S. Kuyzin and K. A. Lemieux, Fundamental Design Equations for Structural RIM, *Proceedings of the SPI 34th Annual Technical/Marketing Conference*, Technomic, Lancaster, PA, 1992, pp. 238–245.

23. Standard Methods of Testing: Flexible Cellular Materials—Slabstock, Bonded and Molded Urethane Foams, American Society for Testing and Materials, D3574-86; ASTM, Philadelphia, 1986.

24. T. Yamamoto, M. Shibayama, and S. Nomura, *Polym. J. 11*:895 (1989).

25. H. W. Wolfe, Cushioning and Fatigue, *Mechanics of Cellular Plastics* (N.C. Hilyard, Ed.), Macmillan, New York, 1979, p. 99.

26. R. P. Kane, *J. Cell. Plast. 1*:217 (1965).

27. R. M. Herrington and D. L. Klarfeld, *J. Cell. Plast. 20*:58 (1984).

28. R. G. Skorpenske, R. Solis, A. K. Schrock, and R. B. Turner, Compression Set Mechanisms in Flexible Polyurethane Foam, *Proceedings of the SPI 34th Annual Technical/Marketing Conference*, Technomic, Lancaster, PA, 1992, pp. 650–659.

29. L. J. Gibson and M. F. Ashby, *Cellular Solids: Structure and Properties*, Pergamon Press, Oxford, 1988.

30. W. M. Lee, Cell Structure Deformation of High Resilient Polyurethane Foam by Scanning Electron Microscopy, *Proceedings of the SPI 6th International Conference of the Polyurethane Division*, Technomic, Lancaster, PA, 1983, pp. 422–427.

31. C. Kau, L. Huber, A. Hiltner, and E. Baer, *J. Appl. Polym. Sci. 44*:2069, 2081 (1992).

32. E. H. McKenna, S. E. Wujcik, and R. F. Pask, Improved Fatigue Resistance of Polyurethane Carpet Underlay, *Proceedings of the SPI 33rd Annual Technical/Marketing Conference*, Technomic, Lancaster, PA, 1990, pp. 473–479.

33. A. N. Gent and A. G. Thomas, *J. Appl. Polym. Sci. 1*:107 (1959).
34. N. C. Hilyard, *Mechanics of Cellular Plastics*, Macmillan, New York, 1979.
35. K. C. Rusch, *J. Appl. Polym. Sci. 10*:2299 (1969).
36. A. Nagy, W. L. Ko, and U. S. Lindholm, *J. Cell. Plast. 10*:127 (1974).
37. B. P. Parma, M. B. Rhodes, and R. Salovey, *J. Appl. Phys. 49*:4985 (1985).
38. F. J. Dwyer, *J. Cell. Plast. 12*:104 (1976).
39. H. W. Wolfe, D. F. Brizzolara, and J. D. Bryam, *J. Cell. Plast. 13*:48 (1977).
40. J. Hoffmann, H. Ostromow, F. Prager, H. M. Rothermel, and J. Vogel, Determination of the Composition and Properties of Polyurethanes, *Polyurethane Handbook* (G. Oertel, Ed.), Hanser Publishers, New York, 1985, pp. 450–494.
41. G. Woods, *Flexible Polyurethane Foams: Chemistry and Technology*, Applied Science Publishers, London, 1982, pp. 126–128.
42. *Fatigue of Virgin Urethane Carpet Cushion*, Product Bulletin SC-788, Union Carbide Urethane Intermediates Division, 1988.
43. B. Beals, F. J. Dwyer, and M. Kaplan, *J. Cell. Plast. 1*:32 (1965).
44. G. W. Ball and D. J. Doherty, *J. Cell. Plast. 3*:223 (1967).
45. R. E. Jones and G. Fesman, *J. Cell. Plast. 1*:200–216 (1965).
46. H. Stone, Fatigue Testing of Flexible Foams, *Proceedings of the SPI 27th Annual Technical/Marketing Conference*, Technomic, Lancaster, PA, 1982, pp. 124–137.
47. W. A. Ashe, Fatigue Test for Carpet Cushion, *Proceedings of the SPI 30th Annual Technical/Marketing Conference*, Technomic, Lancaster, PA, 1986, pp. 320–325.
48. K. D. Cavender, New Dynamic Flex Durability Test, *Proceedings of the SPI 33rd Annual Technical/Marketing Conference*, Technomic, Lancaster, PA, 1990, pp. 282–288.
49. B. N. Stevens, J. F. Scott, D. J. Burchell, and F. O. Baskent, *J. Cell. Plast. 26*:19 (1990).
50. S. M. Clift, Understanding of Dynamic Properties of Polyurethane Cast Elastomers, *Proceedings of the SPI 33rd Annual Technical/Marketing Conference*, Technomic, Lancaster, PA, 1990, pp. 547–553.
51. H. J. Kogelnik, H. H. Huang, M. Barnes, and R. Meichner, Comparison of the Dynamic Properties of Solid Polyurethane

Elastomers, *Proceedings of the SPI 33rd Annual Technical/ Marketing Conference*, Technomic, Lancaster, PA, 1990, pp. 207–219.

52. J. D. Ferry, *Viscoelastic Properties of Polymers*, John Wiley, New York, 1961.

53. L. E. Neilsen, *Mechanical Properties of Polymers and Composites*, Marcel Dekker, New York, 1974.

54. C. S. P. Sung, T. W. Smith, C. B. Hu, and N. H. Sung, *Macromolecules 12*:538 (1979).

55. J. L. Work, *Macromolecules 9*:759 (1976).

56. W. Nierzwicki and E. Wycocka, *J. Appl. Polym. Sci. 25*:739 (1980).

57. B. Hartmann, J. V. Duffy, C. F. Lee, and E. Balizer, *J. Appl. Polym. Sci. 37*:1829 (1988).

58. C. B. Hu, R. S. Wand, and N. S. Schneider, *J. Appl. Polym. Sci. 27*:2167 (1982).

59. K. K. S. Hwang, S. B. Lin, S. Y. Tsay, and S. L. Cooper, *Polymer 25*:947 (1984).

60. J. W. Dieter and C. A. Byrne, *Polym. Eng. Sci. 27*:673 (1987).

61. E. C. Prolingheuer, J. J. Lindsey, and H. Kleimann, *J. Elastomers Plast. 21*:100 (1989).

62. R. B. Bird, W. E. Stewart, and E. W. Lightfoot, *Transport Phenomena*, John Wiley, New York, 1960.

63. J. W. Reisch and D. M. Capone, Polyurethane Sealants and Cast Elastomers with Superior Physical Properties, *Proceedings of the SPI 33rd Annual Technical/Marketing Conference*, Technomic, Lancaster, PA, 1990, pp. 368–374.

64. C. P. Christenson, M. A. Harthcock, M. D. Meadows, H. L. Spell, and R. B. Turner, *J. Polym. Sci. 24*:1401 (1986).

10
Design of Tough Epoxy Thermosets

BRUCE L. BURTON and
JAMES L. BERTRAM*   The Dow Chemical
Company, Freeport, Texas

I. INTRODUCTION

The desirable, high glass transition temperatures of epoxy thermosets are largely attributed to their crosslinked state. Unfortunately, high glass transition temperatures are usually obtained at the expense of reduced toughness and damage tolerance caused by the highly crosslinked structure. The combination of high glass transition temperature and good damage tolerance is particularly desirable for high-performance composite applications.

New ways of increasing toughness while retaining high-temperature performance (e.g., T_g and moduli) have been sought for over two decades. The bulk of this research has focused on epoxy resin systems for applications such as com-

---

*Retired

posites or adhesives used in aerospace or electronic markets. Historically, the most widely used means of improving the toughness of epoxy resin systems has been through the addition of resin soluble rubbers. Depending upon the type of rubber and epoxy resin system used, the rubber may form a secondary phase during the epoxide polymerization reaction. The degree to which this phase separation occurs can control not only the amount of toughening obtained, but also a number of other properties such as the glass transition temperature and the modulus of the cured (i.e., polymerized) system.

"Tougher" is a relative term used to describe a material that requires greater stress or energy levels to reach a defined "failure point." The definition of failure itself may vary considerably depending upon the application and/or material involved in the test. Thus, "toughness" may or may not be a material parameter, depending upon which definition of failure is chosen. Consequently, for many composite materials, where toughness is vigorously sought, the toughness rankings for several resin systems may vary with the application and test method. This has been a cause of confusion.

Although there are many facets to designing a tough, fiber-reinforced composite, the use of a tough resin is axiomatic. As tougher and tougher composites are developed, the focus of the problem may repeatedly shift between an emphasis on the matrix resin and an emphasis on other items affecting the composite such as the resin fiber "interphase," the three-dimensional architecture of the fibers, second-phase particulate toughening, or the fiber itself. As the first step in developing tougher composite materials a study was made of neat resin castings at low strain rates in order to better define the relationship between "toughness" and cured resin structure. As described later, special attention was paid to the effect of decreasing the number of crosslinks in the cured resin system.

It is generally accepted by those who work with thermosetting polymers that as the glass transition temperature (T_g) of the polymer increases, it becomes more brittle. Thus, a simple means of toughening is to lower the T_g of the system as far as possible (within the constraints of the application's requirements). This can be done by using reactants that will provide more chain flexibility in the polymer or by the addition of nonreactive plasticizers.

II. YIELD STRENGTH

Yielding refers to the onset of plastic deformation in a material as it is loaded. The yield stress of crosslinked polymers is closely related to their toughness [1–3]. It was found that the compressive yield strengths of a wide variety of polymerized epoxy resin systems were positively correlated with glass transition temperature (measured at 10°C/min via a thermomechanical analyzer) as shown in Fig. 1. Those systems exhibiting glass

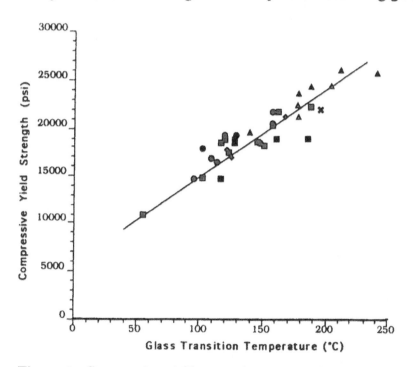

Figure 1 Compressive yield strength versus T_g for various epoxy resin systems. ■, D.E.R.* 383 or 331; ●, D.E.N.* 431; ♦, D.E.R.* 337; ▲, D.E.N.* 438; ✗, Tactix* 742. (*Trademark of The Dow Chemical Company.)

transition temperatures less than about 160°C underwent ductile yielding at a creep rate of 0.002 in./min, whereas the cured resin systems with T_gs greater than about 160°C underwent brittle fracture prior to reaching this creep rate. Similar behavior was observed for several vinyl ester resin systems.

In Fig. 2 it is seen that compressive yield strength versus T_g data for several polycarbonates [4] (a good representative of a "tough" polymer) lie on a "line" below that of the cured epoxy resins. Upon physical aging, the yield stresses of these polycarbonates were seen to increase, as indicated by the direction of the dashed line. Polycarbonates are known to become more brittle as they physically age. Figures 1 and 2 imply that

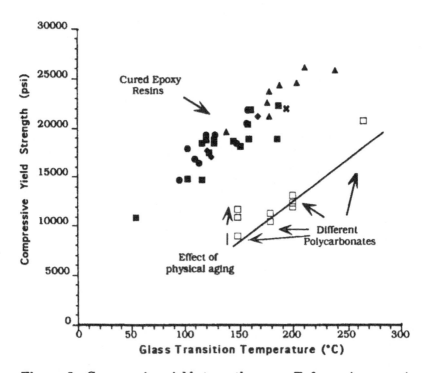

Figure 2   Compressive yield strength versus T_g for various epoxies and polycarbonates.

in order to obtain tougher thermosets, it is necessary to develop resins that exhibit lower yield stresses.

Though positive correlations have been reported between yield stress and T_g [5], relatively little appears to have been published on the relationship of T_g to changes in network structure caused by physical aging. If physical aging readily occurs near the use temperature of a resin system, it may eventually become more brittle than systems that were previously tougher. Thus, the "toughest" system should be determined by using specimens that have been aged under actual service conditions. Truong and Ennis [6] state that the "increased yield stress after physical aging is the main factor contributing to the reduction in fracture toughness."

III. EFFECTS OF RUBBER

Figure 3 shows compressive yield stress versus T_g data for some rubber-modified epoxies superimposed on the lines generated from the data of Fig. 2. Increasing levels (5, 10, and 15 wt %) of a given Hycar* CTBN rubber gave increasing amounts of yield stress reduction, with slight decreases in glass transition temperature. A relatively greater yield stress reduction is seen for the same resin system, modified with insoluble polyacrylate rubbers dispersed in an epoxy resin. Several such rubbers were also used to modify D.E.H.[†] 24 (triethylene tetramine) cured D.E.R.[†] 331 epoxy resin. As shown by the "x's" in the figure, most of the data points fell close to one another except for one system in which the rubber showed less phase separation.

Studies of epoxy resins modified with Hycar butadiene-acrylonitrile-based rubbers (BF Goodrich) are by far the most numerous of those published on rubber-modified epoxy resin

---

*Trademark of BF Goodrich Company.

[†]Trademark of The Dow Chemical Company.

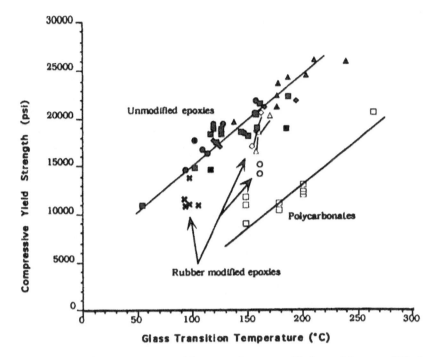

Figure 3 Compressive yield strength versus T_g for rubber-modified epoxy resins. $\Diamond$, D.E.R. 331/MDA/Hycar CTBN 1300X13 (5, 10, and 15 wt %); $\triangle$, D.E.R. 331/MDA/Hycar CTBN 1300X8 (5, 10, & 15 wt %); $\bigcirc$, D.E.R. 331/MDA/epoxy insoluble polyacrylate rubber (16.5 wt %); $\times$, D.E.R. 331/D.E.H. 24/epoxy insoluble polyacrylate rubber (18.1 wt %).

systems [e.g., 7–11]; as described in these and many other papers, polymerization of soluble rubber containing epoxy resin systems causes portions of these rubbers to phase out to varying extents. The composition of the formulated system can greatly affect the amount of rubber that phases out during polymerization. Additionally, the degree of phase separation and the type of rubber morphology formed within a given epoxy system has been shown to depend upon curing temperatures, the prereaction of certain components, and the vitrification or gelation rates of the system [12], although clear trends of morphology development with curing conditions may be lacking [13]. A paper by Verchere et al. [14] provides much informative detail on the complexities of how rubber concentration and polymerization temperature affect the final morphology. The relation of the rubber's phase morphology to cured resin mechanical properties has been studied for several resin systems but many of these, such as piperidine-cured resins, are model systems of limited practical utility [15].

It is widely held that because of the low thermal conductivity of epoxy resin systems, the heat of polymerization will cause temperature differences within molded parts of variable thickness, leading to variations in the rubber particle size distribution that will result in inconsistent mechanical properties. For many applications, practitioners view such sequential development of the final rubber morphology as more of a potential processing hazard than as a tool to be utilized. Many papers have focused on the correlation of rubber particle size distribution to "toughness." Since toughness itself is a fuzzy concept, related in practice to the specific applications, it is little wonder that the literature often appears filled with "conflicting" results. Whether or not bimodal particle distributions enhance toughness is a good example of this phenomenon [16,17]. The correct answer appears to be that sometimes they do and sometimes they don't. Although several papers describe how cure times and temperatures affect morphology, Grillet et al. [17] recently studied systems in which the cure schedule had no clear effect on morphology. They also state that the "principal parameter for the toughness of rubber-modified epoxies is the volume fraction and not the diameter of the dispersed phase particles." Another researcher has focused on the interparticle ligament thickness of the matrix as being the key to toughening in these systems [18].

For some applications, a serious drawback to using rubbers that form a secondary phase is the modulus reduction they cause. Additionally, modulus can vary with morphology as do T_g and toughness. The modulus of rubber-modified materials typically deviates from "rule of mixtures" predictions, having somewhat higher than expected values at low rubber levels and progressively lower than expected values as the rubber phase begins to percolate through the matrix. Although such modulus reduction may itself contribute to toughness, since stresses don't rise as high at a given deformation level (especially important where stress-concentrating impurities may be present), most applications require some minimum level of stiffness which is often best gained via high modulus materials since they can provide weight and volume savings. As discussed later, polymerized epoxy resin systems often have higher moduli as their glass transition temperatures are reduced. The reduction in yield stress upon going to lower T_g systems appears to provide increased ductility, in spite of the resin's higher modulus.

Referring again to Fig. 3, rubbers that remain completely soluble in the cured matrix may also exhibit lower yield stresses but are expected to do so by moving downward along a line parallel to that shown for the unmodified epoxy systems. Thus, if toughness must be increased without decreasing T_g, the amount of rubber that remains dissolved in the matrix should be minimized. For this reason, a rubber that remains insoluble in the resinous reaction mixture throughout the polymerization process, such as an insoluble polyacrylate dispersion, will exhibit the most consistent toughening (under varying cure conditions) and exhibit the smallest decrease in glass transition temperature.

IV.   RUBBER DISPERSIONS IN EPOXY RESIN

Several rubber modifiers were used in these studies that were available from internal laboratories. These insoluble rubbers made investigation of toughening effects possible because there was no ambiguity about which phase the toughening agent was in. Preformed dispersions of epoxy insoluble rubbers [19] are particularly useful in many adhesive and composite applications because the rubbery-phase volume of the final plastic is relatively insensitive to variations in cure temperature. Consequently, rubber levels that provide optimum toughening may be more easily formulated for a given application. In addition, the modulus of materials made with such rubbers can be varied independently of T_g since the T_g does not decrease with increasing rubber content.

An experimental rubber-modified epoxy resin was prepared by making a 40% dispersion of epoxy-insoluble polyacrylate rubber in Tactix* 123 epoxy resin (predominantly the diglycidyl ether of bisphenol A) [20,21]. Such high rubber content provides increased latitude in formulating. Using this product, one may add significant amounts of rubber to a formulation without decreasing its T_g. Unlike resins that contain soluble rubbers, this product is easy to handle when warm and does not draw out in the long, stringy fashion commonly occurring in many soluble-rubber-containing resins. Additional details on the mechanical properties of cured resins incorporating this rubber, and others, have been described by Hoffman and Kolb [20] and Sue et al. [22,23].

Another method of preparing rubber-modifying resin systems is to blend liquid-dispersed rubbers into one of the system's components and then remove the dispersing liquid by distillation. This method [24] is detailed by D. Pickelman in Chapter 3.

V. TOUGH EPOXY RESINS—WITHOUT RUBBER

Would matrices be toughened if one could remove the rubber from a two-phase system, leaving only holes? Work shedding some light on the answer was initially reported by Waddill [25] and has recently been further investigated by Huang and

---

*Trademark of The Dow Chemical Company.

Kinloch [26]. Microvoids created using Jeffamine™* BuD 2000 (a urea-terminated polyether amine) were found to toughen a relatively ductile epoxy system significantly. As expected, the yield stress was reduced, although the toughening was not as great as that found in some rubber-containing resins because of the absence of some of the energy-absorbing failure modes (e.g., crack bridging by rubber particles). Such systems might well serve to further elucidate the concepts developed by Wu where the criterion for rubber toughening of resins is described in terms of the "critical matrix ligament thickness" [18] that exists between the rubber particles. Fukui et al. [27] presented a theoretical analysis of rubber-modified nylon predicting that the stress concentrations would be more uniform in the rubber-containing system than in the void-containing system. The void-containing system was predicted to have slightly less yielding than the rubber-containing system and fractured elements appeared at lower stresses in the void-containing system. This suggests that the strain at break would be larger for the rubber-containing system.

In considering Figs. 1 and 2, focus on what appear to be the main structural distinctions between the polycarbonates and the brittle, unmodified epoxy resins. Polycarbonate, which shows a desirable yield stress/T_g relationship, has polymer chains that are (1) more loosely "crosslinked" or entangled, (2) less polar, and (3) stiffer than the conventional epoxy resin systems studied. The effects of these differences are discussed below.

VI. CROSSLINK DENSITY/CHAIN STIFFNESS

In some applications areas, increased toughness and T_g are simultaneously desired. Since increased crosslinking tends to

---

*Trademark of Huntsman Corporation.

increase a polymer's T_g but decreases its toughenability [28], ways were sought to maintain high glass transition temperatures while reducing crosslink density, thus increasing polymers' toughnes at a given T_g. Since T_g depends upon the "stiffness" of the polymer chains, it should be maintained at reduced crosslink densities by (1) increasing the number of noncovalent chain interactions and/or (2) increasing polymer backbone stiffness.

The molecular weight between crosslinks can be adjusted in epoxy resins by reaction with diphenolic compounds. The ratio of diphenolic to epoxy determines the degree of advancement. The ratio of the diphenolic advancement component to curing agent determines the molecular weight between crosslinks. Typically, with resin advancement using bisphenol A, as the crosslink density increased (i.e., lower molecular weight between crosslinks), the toughness decreased. However when stiff diphenols, relative to bisphenol A (e.g., tetraboromobisphenol A, biphenol, 9,9-bis(4-hydroxyphenyl)fluorene), are used, the T_g is maintained or even increased while the crosslink density of the cured system decreases. The toughness is also increased, eventually changing the failure mode from brittle to ductile.

Somewhat curiously, increasing the stiffness of the polymer backbone need not lead to higher moduli or yield strength values for epoxy thermosets. In recent years it has become increasingly recognized that for epoxy thermosets the relationship between the glass transition temperature and modulus often runs counter to that generally reported for thermoplastic polymers. Specifically, the modulus of a wide variety of epoxy resin systems has been seen to decrease as their T_g's increase. This behavior has been studied in detail by Gillham and students [29–33]. It is believed that the density decreases with increasing T_g (due to increased crosslinking) because the increase in unoccupied volume that is frozen into the glass, by virtue of the higher glass transition temperature, is greater than the decrease in unoccupied volume that is caused by polymerization shrinkage at the curing temperature. Similarly, the moduli of specific resin systems have also been shown to decrease as further curing occurs in undercured (incompletely polymerized) systems [34–36]. Kong et al. [37] showed that D.E.R.* 332 epoxy resin cured with different quantities of Jeffamine† T403 (a polyether triamine) had a minimum density (maximum specific volume and compliance) at a one-to-one stoichiometry.

VII. POLYMER NETWORK ARCHITECTURE

A. Molecular Scale

Quite often, to obtain increases in T_g, reaction mixtures of higher average functionality are used to obtain higher crosslink densities. As the average functionality of the reactants is increased, the mixtures will gel or vitrify at lower and lower percentages of "complete" reaction [38]. As a result, higher and higher temperatures must then be used to force the polymerization reaction to completion. For epoxies, a greater number of epoxide-hydroxyl side reactions occur at these higher temperatures with the result that the network become less and less homogeneous, with their final structure being dependent upon the cure temperature schedule. Network defects may also be introduced via degradation reactions that may occur at elevated temperatures.

Final polymer network architecture may also be varied by addition order or prereaction of certain components. The studies of the effects of network architecture changes, that is, changes in the distribution of the network chains in systems of initially constant chemical composition, have shown interesting results that can differ for various systems. It has been observed that of several ultimate properties studied, only tensile

---

*Trademark of The Dow Chemical Company.
†Trademark of Huntsman Corporation.

strength and the energy under the ambient stress–strain curve were influenced by variation in the crosslink density distribution, suggesting that the effects could be due to either (1) concentration of stress in the short chains (Bueche's theory [39]), or (2) changes in structure due to viscosity differences and/or phase clustering during curing. Misra et al. [40] reported that for the cured epoxy systems in their study, changing the distribution of M_c (molecular weight between crosslinks) affects T_g (sometimes), the slope of the transition region (dynamic mechanical spectroscopy), the shape of the tan δ peak, and, especially, the characteristic creep time of the polymer. No changes were found, with changing M_c distribution, for rubbery modulus, density, room temperature impact strength and tensile properties, and β-transition temperature. Kim et al. [41] reported additional details on the creep studies. Wu et al. [42] reported that systems made to have more heterogeneous distributions of crosslinks showed no variation in T_g or modulus but did show increases in fracture toughness and energy, by factors of 2 and 4 (respectively). Wu et al. achieved these variations via prereaction of certain components, but variations of initial cure temperature in a given system have also been found to produce different network structures [43].

Greater network homogeneity (in terms of eliminating unreacted chain ends), which is likely to result from gelling epoxy systems at higher conversion percentages, may be one of the keys to toughening high T_g epoxy systems. The importance of network homogeneity may be viewed as follows: If a network is quite homogeneous, that is, if the chain lengths between junctions are of similar length, more chains will bear the load when stressed than if the same number of chains were present in a less homogeneous network. If the less homogeneous network is tested in tension, a smaller number of shorter chains or network segments, which bear the load first, will break and shift the load to the next shortest group of chains. This process will repeat itself until it propagates so quickly that macroscopic failure is observed.

The shorter chains in the less homogeneous network may bear as much stress, individually, before failure, as the chains in the homogeneous network, but because there are fewer of them, the macroscopic load at failure will be lower than in the homogeneous network. The ultimate strain at failure will also decrease because the chains or segments that might be expected to yield greater elongations at break experience, microscopically, increasingly greater loading rates. Macroscopic failure occurs because the localized loading rate has now become faster than the fastest rate at which the remaining chains can undergo conformational changes that will relieve localized stress concentrations. Thus, using formulations of low average functionality, without the presence of monofunctional chain-terminating species, may allow formation of more homogeneous networks with improved toughness related properties.

From a practical standpoint however, there are potential benefits from gelling the systems at lower, rather than higher, total epoxide conversions. Such benefits are discussed by Riccardi and Williams [44], who developed and studied partially prepolymerized resin systems that gel at relatively low conversions and suggest the use of this methodology as a means of decreasing demolding times.

B. Microscopic Scale: Nodules and Mixing

Much has been written about the existence, nonexistence, importance, irrelevance, and cause of what are often referred to as "nodules" in cured epoxy resin systems [42,45–47]. Several of these papers describe correlations between the morphology of the cured resins and their mechanical properties, suggesting that such properties may be optimized through a fundamental morphological control. Similar-appearing nodules have also been seen in polycarbonates [48]. It has been reported [49] that the nodules' appearance in, and the T_g and tensile properties of, polymerized epoxy resins may be changed as a result of changes in the method used for mixing. It has also been found that improved mixing can increase the T_g of cured resins without affecting their tensile strength. For the sake of reproducible data, optimum mixing should be obtained, whatever its effect on nodules and toughness or strength. Heterogeneous networks containing "microgels" and characterized by multiple glass transition temperatures have also been reported [50]. It has not been proved whether such microgels are related to "nodules."

VIII. POLARITY

Most epoxy resins are cured via addition reactions with active hydrogen-containing compounds at close to a one-to-one stoichiometry, rather than via "catalytic" curing agents (such as small amounts of Lewis acids or bases). Because of this, crosslinking and polarity are closely related in most epoxy resin systems. When higher T_g's are needed for performance at elevated service temperatures, the most common method for meeting this need has been to reformulate the thermosetting reaction mixtures to give an increased degree of crosslinking. In conventional epoxy resin systems this leads to an increased concentration of secondary hydroxyl groups, which in turn contributes to increased moisture sorption of the resin. Such sorption, which has been seen to range from 1 to 9 wt %, usually has an adverse effect on the physical properties of the cured thermoset by (1) decreasing glass transition temperatures, (2) decreasing moduli, and (3) causing swelling. Swelling causes stresses that can contribute to the failure of epoxy composites. For several different cured epoxy resin systems, it was seen that swelling was a fairly linear function of the amount of water sorbed, even though the systems that sorbed the most water were those expected to have the greatest degree of crosslinking [51].

For the reasons described in the preceding sections, creation of polymers possessing the following characteristics appears desirable:

1. Low polarity, to minimize water absorption and wet T_g loss
2. Lower crosslink density and greater network homogeneity, so that larger numbers of more extensible chain segments could bear any stresses
3. Stiffer polymer chain segments between crosslinks, to maintain the polymers' glass transition temperatures at lower crosslink densities.

This approach lead to the development of polymers that have been nicknamed "CET" resins (vide infra) for crosslinkable epoxy thermoplastics.

A.   CET Resins: Tactix* 695 Epoxy Resin and XUS 19020.00 Experimental Resin

CET resins consist of a mixture of epoxy resins that also contain difunctional extender materials (typically diphenols), a multifunctional, dual active curing agent that is capable of linear advancement prior to crosslinking, an additional curing agent (optionally), and a latent catalyst to promote the advancement reaction of the epoxy with the diphenol [52,53]. When cured, these resins appear to be single phase. It has been hypothesized that the improvement in toughness is due to the ability of these systems to generate relatively high molecular weight, uncrosslinked segments that contain crosslinkable sites. Toward the end of the polymerization, these sites help create a high degree of "network perfection." A large network mesh size is expected of this cured resin system since the average functionality of the uncured mixture is low, as is

*Trademark of The Dow Chemical Company.

the temperature throughout much of the curing time. (See Table 1 for a comparison of in situ advancement versus step advancement/crosslinking). Note that at equal T_g's, the toughness of the CET system is approximately double that of the conventional system at identical crosslink densities. It is further speculated that the improved toughness/T_g combination of these resins is the result of a higher fractional free-volume content resulting from greater chain stiffness than conventional resins [54]. Morgan [55] states that the "flexibility and extensibility of a crosslinked epoxy network are determined by the available glassy-state free volume," and Morgan as well as others have shown correlations between the glassy-state free volume and the yield stress of cured resins.

Examples 1 and 2 in Table 1 are CET-type epoxy resin systems composed of the same reactants that were processed exactly the same way, differing only in reactant ratios. Comparative examples A and B are equivalent to examples 1 and 2, respectively, except that certain components of the formulation were mixed and heated at 150°C for 1 hour prior to addi-

Table 1 Effect of In Situ Advancement on Cured Epoxy Resin Properties

Cured properties	Example no. 1	Comparative example A	Example no. 2	Comparative example B
T_g (°C)	141.4	136.8	126.7	127.2
G_{IC} (kJ/m^2)	1.51	1.10	1.62	1.06
Izod impact (ft-lb/in.) (unnotched)	14.3	9.7	34.8	20.5
Elongation (%)	9.5	5.3	19.5	10.8
Uncured viscosity (cps at 100°C)	80	>4000	80	>4000

tion of the remaining components. The mixture was then cured as in examples 1 and 2.

Bertram et al. showed [53] that when some classes of polymerized thermosets were examined over a sufficiently broad range of T_g, plots of strain energy release rate (G_{1C}) versus glass transition temperature took on a somewhat hyperbolic shape (Fig. 4). CET-type resin systems and two-phase

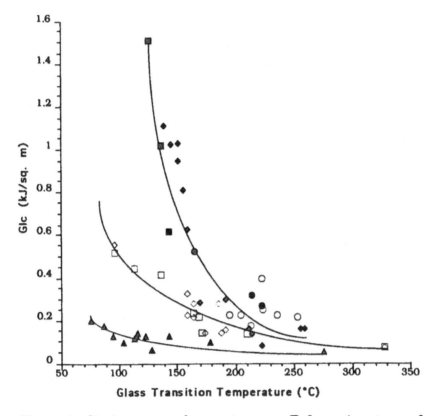

Figure 4   Strain energy release rate versus T_g for various types of epoxy resin systems. □, conventional resins; ■, CET-type resins; ◇, Tactix 742 based; ◆, rubber-modified Tactix 742; ○, Tactix 742 based CETs; ●, rubber-modified Tactix 742 based CET-type resins; ▲, TGMDA/various hardeners.

rubber-modified systems yielded data that was shifted upward and to the right of untoughened epoxy systems, indicating that the modified systems are tougher at a given glass transition temperature. For T_g's in excess of 200°C, C.E.T.-type systems were the toughest observed, based on strain energy release rate data.

Table 2 compares the thermal and mechanical properties of three different types of CET resins with a typical high-performance epoxy system. The same basic technology is used to produce each of the three CET resins. Other experimental CET products have been made that exhibit higher T_g's while

Table 2   Property Comparison of Some Clear Cast Resins

Property	Hardener Added →	XUS 19020.00 none—1 component	Tactix 695 DDS	Exp. CET type resin DDS	TGMDA DDS
Tensile					
Strength (psi)		13.0	13.2	12.0	8.5
Modulus (ksi)		441	444	420	540
Elongation at break (%)		5.4	8.3	8.0	1.8
Flexural					
Strength (psi)		21.1	19.1	20.0	19.9
Modulus (ksi)		496	456	505	585
% strain		>5%	>5%	>5%	—
G_{1C} (strain energy release rate)					
Compact tension (kJ/m²)		0.24	0.14	—	0.07
Compact tension (in.-lb/in.²)		1.40	0.82	—	0.40
Double torsion (in.-lb/in²)		—	1.00	2.45	—
Water absorption (wt %)					
200 hr in boiling water		—	1.25	0.90	6.00
2 weeks in boiling water		1.41	1.62	—	—
Glass transition temperature (°C by DMA)					
Dry		155	165	145	238
Wet (200 hr in boiling water)		—	150	135	120
Density (g/cc)		1.35	1.48	1.53	—

still retaining good toughness. The low water absorption of the CET resins and their excellent retention of T_g while wet are shown at the bottom of Table 2. Further information on the thermal properties of these resins is given elsewhere [54].

Graphite fiber composite data is presented in Table 3. High compressive strength, especially after impact damage, is a design critical parameter for aircraft structures. The laminate used in testing compression after impact was a 40 ply quasi-isotropic layup (0, +/−45, 90° fiber orientation). Tactix 695 epoxy resin and related experimental CET resins show greater postimpact compression strengths in composites than conventional epoxies with similar T_g's [56]. Additionally, these ductile systems do not exhibit the microcracking seen in composites made from conventional resins.

Since the initial development of these resin systems, another class of toughened epoxies, referred to here as semi-interpenetrating polymer networks or "semi-IPNs" has undergone considerable developmental work by numerous companies and institutions (as a small example, refs. 57–60). These systems are generally prepared by dissolving a thermoplastic (viz., polysulfones, polyethersulfones, polyetherimides) into an epoxy resin and subsequently polymerizing the solution under conditions such that the epoxy resin phases out during curing to form a dispersed, crosslinked second phase. (This is the reverse of systems developed earlier in which a thermoplastic phases out of a continuous epoxy matrix.) In effect, the epoxy portion of the semi-IPN mixture serves as a reactive diluent, which initially provides processability to the thermoplastic but then later solidifies to a rigid filler. Such systems have been the focus of much research, because of their combinations of toughness, high T_g, high modulus (relative to many rubber-modified resins), and processability (relative to thermoplastics), but relatively little data have been reported on such resins for creep behavior, solvent resistance, and stress cracking, which may represent a whole new series of Achilles' heels relative to the thermosetting resins they

Table 3 Properties of Some Carbon-Fiber-Reinforced Composites

Properties	XUS 19020.00 resin, (no hardener—one component resin system)	Tactix 695 resin, DDS hardener	Exp. CET-type resin, DDS hardener	TGMDA resin, DDS hardener
AS4 Carbon Fiber: Unidirectional Laminates				
0° Flexural, 77°F dry				
Strength (ksi)	220			300
Modulus (msi)	17.0			18.5
Strain (%)	1.5			
90° Flexural, 77°F dry				
Strength (ksi)	13			
Modulus (msi)	1.30			
Strain (%)	1.0			
0° Compression—77°F dry by IITRI				
Strength (ksi)	190			250
Modulus (msi)	20.0			20.0
0° Compression strength (ksi)				
Room Temperature, dry		210	180	226
180°F, dry		225	—	—
180°F, wet		170	—	158 (170°F)

Table 3 Continued

Properties	XUS 19020.00 resin, (no hardener—one component resin system)	Tactix 695 resin, DDS hardener	Exp. CET-type resin, DDS hardener	TGMDA resin, DDS hardener
AS4 Carbon Fiber: 40 Ply Quasi-isotropic Layup				
Post-impact compression strength, (ksi)				
1500 in.-lb/in.^{2}		42	50	18
(40 ply quasi-isotropic layup)				
G30-500 8HS Fiber				
0° Flexural, 77°F dry				
Strength (ksi)	152			150
Modulus (msi)	8.0			10.0
Short beam shear strength (ksi)				
77°F Dry	10			
Open-hole compression by IITRI				
Strength (ksi) 77°F, Dry	38			
180°F, wet	34			
Post-impact compression strength (ksi)				
1500 in.-lb/in^2, 77°F, Dry	39			

seek to replace. In the CET resin systems, some of the potential complications associated with the two-phase semi-IPN systems are avoided. In what might be considered a variation on the semi-IPN theme, American Cyanamid Company has done much work on what are termed "interleaf" systems [61–63].

IX. THERMAL EXPANSION BEHAVIOR

In multiphase cured systems the nature and magnitude of the thermal expansion mismatch that occurs between the phases can strongly affect the final properties of the cured material. The inclusion of a secondary phase which has a lower coefficient of thermal expansion (CTE) than the matrix resin (e.g. silica) not only decreases the thermal expansion of the cured matrix resin by "dilution" but also induces a tensile stress on the matrix that serves to further reduce its thermal expansion [64,65]. The magnitude of this effect is related to both (1) the thermal expansion coefficient "mismatch" between the matrix and the filler and (2) the stress state existing at the use temperature.

By analogy, the inclusion of a secondary phase that has a higher coefficient of thermal expansion (CTE) than the matrix resin (e.g., rubber) should induce a compressive stress on the matrix as it cools through T_g. Such a stress should offset the hydrostatic tensile stress present in a cured resin and would therefore increase the ability of the resin to yield in tension.

X. SIMHA–BOYER RELATIONSHIP

Boyer and Simha [66–68] found a correlation between the volume expansion of a polymer's free volume near T_g and the structure of the polymer. The expansion of free volume close to T_g was represented by the difference between the rubbery ($>T_g$) and glassy ($<T_g$) volume thermal expansion coefficients.

Their data on amorphous high polymers often followed the form:

$$(a_r - a_g)T_g = (V_f/V_g) = PT_g = K \tag{1}$$

where a_g is the volume expansion coefficient below T_g (T_g in Kelvin), a_r is the volume expansion coefficient above T_g, and V_f/V_g is the ratio between the frozen-in excess volume caused by molecular immobility in the glassy state (free volume) and V_g, the total volume at the glass transition temperature, T_g [69]. The V_f/V_g, referred to as the *free volume fraction*, was found to be relatively constant for a variety of thermoplastic polymers. See Table 4 for thermal expansion data for various cured epoxy systems.

Bauwens [70] concluded that T_g corresponds to a constant fraction of activated polymer segments and not to a constant value of free volume. He expressed V_f/V_g as PT_g, the fraction of activated polymer segments. This is reasonable since the formation of free volume and the molecular motions giving rise to plastic deformation in the glassy state arise from the same process. Either definition will serve our purposes. Since stress activation of polymer segmental motion may be dependent on the amount of free volume present, it makes sense that polymers with high fractional free volumes at a given T_g are toughest. This may be because less stress-induced dilation is necessary to increase the free volume and/or activate segments of these polymers to levels at which yielding can occur. Many polymers (e.g., epoxies, polycarbonates, PMMA) are known to become more brittle as physical aging decreases their free volume [71]. Taken with Eq. (1), this implies that at a given T_g polymers that have high values of $(a_r - a_g)$ will show the greatest toughness. CET type epoxies, which are unusually tough for their T_g, have higher values of $(a_r - a_g)$ than conventional resins, as shown in Table 4. This corresponds to greater fractional free volumes as shown in Table 5.

Table 4 Thermal Expansion Data for Various Cured Epoxy Systems (in ppm per °C)

Epoxy resin	Epoxy hardener	CLTE at $T > T_g$ (α_r)	CLTE at $T < T_g$ (α_g)	($\alpha_r - \alpha_g$)
Conventional epoxy resin systems				
D.E.R. 331	AEP (aminoethyl piperazine)	179	52.9	126.1
D.E.R. 331	D.E.H.*24 (triethylenetetraamine)	156	62.5	93.5
D.E.R. 331	Bis(p-aminocyclohexyl)methane	129	64.5	64.5
D.E.R. 383	p-menthanediamine	260	69.4	90.6
D.E.R. 383	Ethylenediamine/acrylonitrile adduct	190	79.4	110.6
D.E.R. 383	AEP (aminoethyl piperazine)	282	70.4	211.6
D.E.R. 383	D.E.H. 52 epoxy hardener	156	62.5	93.5
D.E.R. 383	D.E.H. 24 (triethylenetetraamine)	154	76.9	77.1
D.E.R. 383	Bis(p-aminocyclohexyl)methane	145	72.5	72.5
D.E.R. 383	NMA (nadic methyl anhydride)	143	71.4	71.6
D.E.R. 383	BABA (bis(aminobenzyl)aniline)	121	54.5	66.5
D.E.R. 337	bis(p-aminocyclohexyl)methane	135	54.1	80.9
D.E.R. 337	NMA (nadic methyl anhydride)	147	67.2	79.8
D.E.R. 337	BABA (bis(aminobenzyl)aniline)	128	64	64
D.E.N. 431	D.E.H. 24 (triethylenetetraamine)	139	57.7	81.3
D.E.N. 431	D.E.H. 52 epoxy hardener	121	53.8	67.2
D.E.N. 431	Bis(p-aminocyclohexyl)methane	145	48.3	96.7
D.E.N. 431	BABA (bis(aminobenzyl)aniline)	125	55.6	69.4
D.E.N. 431	NMA (nadic methyl anhydride)	122	61	61
D.E.N. 431	AEP (aminoethyl piperazine)	167	55.6	111.4
D.E.N. 438	Ethylenediamine/acrylonitrile adduct	175	58.5	116.5
D.E.N. 438	Bis(p-aminocyclohexyl)methane	96	58	38.6
D.E.N. 438	NMA (nadic methyl anhydride)	125	57.1	67.9

Table 4 Continued

Epoxy resin	Epoxy hardener	CLTE at $T > T_g$ (α_r)	CLTE at $T < T_g$	
			(α_g)	$(\alpha_r - \alpha_g)$
D.E.N. 438	BABA (bis(aminobenzyl)aniline)	87	54.3	32.7
D.E.N. 438	MDA (methylene dianiline)	166	77.7	89
CET-type epoxy resin systems				
Tactix*695	DDS (diaminodiphenylsulfone)	177	56.7	120.4
Tactix*695	DDS (diaminodiphenylsulfone)	187	62.9	124.4
Tactix*695	DDS (diaminodiphenylsulfone)	196	69.8	127.1
Tactix*695	DDS (diaminodiphenylsulfone)	182	61.2	120.8
XUS 19020.00	None added, one-component system	204	70	134
XUS 19020.00	None added, one-component system	194	70.4	123.6
XUS 19020.00	None added, one-component system	187	68.4	118.6
"CET 20A"	None added, one-component system	188	60.8	127.2
"CET 18B"	None added, one-component system	173	70.8	102.2
"CET 14C"	None added, one-component system	157	75.2	81.8
Thermoplastic				
PS	Polystyrene	159	57.3	101.7
Tyril*	Poly(37% styrene–27% acrylonitrile)	158	74.7	83.3
SEMI	Poly(70% styrene–30% ethylmaleimide)	134.7	54.3	80.4
CL STY	Polychlorostyrene (mixed isomers)	126.7	38	88.7
SMI	Poly(75% styrene–25% maleimide)	163.3	61.3	102
T BU STY	Poly(t-butyl styrene)	194	82.3	111.7
PMMA	Polymethyl methacrylate	166	89	77
PVT	Polyvinyltoluene (mixed isomers)	123	54.3	68.7
Zerlon*	Poly(33% styrene–67% methyl methacrylate)	155	64.3	90.7

*Trademark of The Dow Chemical Company.

Table 5 T_g and Compressive Yield Stress Data for Cured Epoxy Systems

Epoxy resin	Epoxy hardener	T_g (°C)	Fractional free volume (V_f/V_t)	Compressive yield strength (psi)
Conventional epoxy resin systems				
D.E.R. 331	AEP (aminoethyl piperazine)	111	0.145	14,250
D.E.R. 331	D.E.H.* 24 (triethylenetetraamine)	117	0.109	—
D.E.R. 331	Bis(p-aminocyclohexyl)methane	150	0.082	18,250
D.E.R. 383	p-menthanediamine	52	0.186	21,440
D.E.R. 383	Ethylenediamine/acrylonitrile adduct	51	0.108	—
D.E.R. 383	AEP (aminoethyl piperazine)	102	0.238	14,500
D.E.R. 383	D.E.H. 52 epoxy hardener	117	0.109	18,000
D.E.R. 383	D.E.H. 24 (triethylenetetraamine)	122	0.091	18,500
D.E.R. 383	Bis(p-aminocyclohexyl)methane	127	0.087	18,500
D.E.R. 383	NMA (nadic methyl anhydride)	142	0.089	18,600
D.E.R. 383	BABA (bis(aminobenzyl)aniline)	187	0.092	21,906
D.E.R. 337	bis(p-aminocyclohexyl)methane	126	0.097	16,750
D.E.R. 337	NMA (nadic methyl anhydride)	152	0.102	—
D.E.R. 337	BABA (bis(aminobenzyl)aniline)	152	0.082	20,870
D.E.N. 431	D.E.H. 24 (triethylenetetraamine)	128	0.098	18,970
D.E.N. 431	D.E.H. 52 epoxy hardener	107	0.077	16,500
D.E.N. 431	Bis(p-aminocyclohexyl)methane	147	0.122	18,100
D.E.N. 431	BABA (bis(aminobenzyl)aniline)	157	0.090	—
D.E.N. 431	NMA (nadic methyl anhydride)	157	0.079	20,210
D.E.N. 431	AEP (aminoethyl piperazine)	112	0.129	16,100
D.E.N. 438	ethylenediamine/acrylonitrile adduct	55	0.115	—
D.E.N. 438	Bis(p-aminocyclohexyl)methane	202	0.055	—
D.E.N. 438	NMA (nadic methyl anhydride)	185	0.093	—

Table 5 Continued

Epoxy resin	Epoxy hardener	T_g (°C)	Fractional free volume (V_f/V_t)	Compressive yield strength (psi)
D.E.N. 438	BABA (bis(aminobenzyl)aniline)	205	0.047	24,100
D.E.N. 438	MDA (methylene dianiline)	150	0.113	—
CET-type epoxy resin systems				
Tactix*695	DDS (diaminodiphenylsulfone)	157	0.155	—
Tactix*695	DDS (diaminodiphenylsulfone)	153	0.159	—
Tactix*695	DDS (diaminodiphenylsulfone)	149	0.161	—
Tactix*695	DDS (diaminodiphenylsulfone)	153	0.154	—
XUS 19020.00	None added, one-component system	138	0.165	—
XUS 19020.00	None added, one-component system	138	0.152	—
XUS 19020.00	None added, one-component system	138	0.146	—
"CET 20A"	None added, one-component system	159	0.165	—
"CET 18B"	None added, one-component system	176	0.138	—
"CET 14C"	None added, one-component system	185	0.112	—
Thermoplastic				
PS	Polystyrene	97	0.113	—
Tyril*	Poly(37% styrene–27% acrylonitrile)	111	0.960	—
SEMI	Poly(70% styrene–30% ethylmaleimide)	131	0.069	—
CL STY	Polychlorostyrene (mixed isomers)	116	0.104	—
SMI	Poly(75% styrene–25% maleimide)	172	0.136	—
T BU STY	Poly(t-butyl styrene)	132	0.134	—
PMMA	Polymethyl methacrylate	96	0.085	—
PVT	Polyvinyltoluene (mixed isomers)	80	0.073	—
Zerlon®	Poly(33% styrene–67% Methyl methacrylate)	104	0.103	—

*Trademark of The Dow Chemical Company.

XI. THERMAL EXPANSION MEASUREMENTS

A. Cured Epoxy Resin Systems

Figure 5 is a plot of the rubbery ($>T_g$) and glassy ($<T_g$) coefficients of linear thermal expansion (CLTEs) versus T_g for a wide variety of cured epoxy resin systems. (Although CLTEs

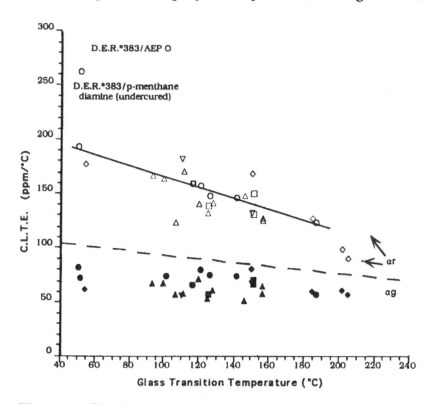

Figure 5 The linear thermal expansion coefficients of various cured epoxy resin systems. Hollow symbols denote a_r (CLTE at temperatures $> T_g$), solid symbols denote a_g (CLTE at temperatures $< T_g$); *epoxy resin types (various hardeners were used)*. ◯, D.E.R. 383; △, D.E.N. 431 or DGE bisphenol F; ▽, D.E.R. 331; □, D.E.R. 337; ◇, D.E.N. 438.

were measured here, the volume expansion coefficient is closely approximated by three times the CLTE [72]; thus the theory above may still be applied here.) It is seen that rubbery CLTEs tended to decrease with increasing T_g. The glassy CLTEs, however, remained nearly constant as T_g increased. These observations indicate:

1. T_g's of cured epoxy resin cannot be measured above a certain critical temperature (about 270°C) using the thermomechanical analyzer (TMA) because the crosslink density becomes so great that the polymer does not expand any faster as a rubber than it does as a glass.
2. The decrease in a_r and toughness found with increasing T_g are both related to the flexibility of the polymer chain. At high crosslink densities, which appear to correspond to low a_r, bond rotation induced by stress causes little increase in ultimate strain.

This last point is particularly important because it means that it should be possible to measure the relative toughness of a cured resin by comparing the difference between the rubbery and glassy CLTEs at a given T_g. According to Eq. (1), systems that are "toughest" (at a given T_g) should be those in which the "fractional free volume" or the "fraction of activated segments" (V_f/V_g or PT_g) is greatest, that is, greatest ($a_r - a_g$). This is easily determined. If TMA plots of cured resins (run at the same heating rate) are compared on the same scale, the tougher system should be that which shows the smallest angle (<180°) between the glassy and rubbery expansion curves (Fig. 6). This is because the smallest angle will correspond to the largest value of ($a_r - a_g$).

Although the data points for both the rubbery and glassy CLTEs show considerable spread at any given T_g, the linearity of most of the data points plotted for D.E.R.* 383 epoxy resin

---

*Trademark of The Dow Chemical Company.

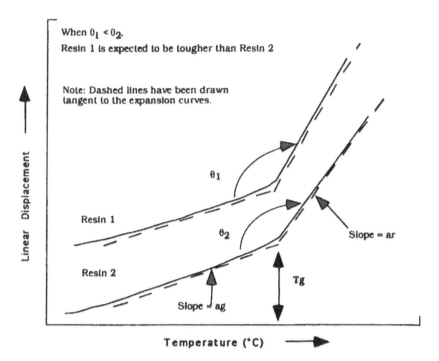

Figure 6 An idealized comparison of thermomechanical analyzer curves for two different epoxy resins. When $\theta_1 < \theta_2$, resin 1 is expected to be tougher than resin 2. Dashed lines have been drawn tangent to the expansion curves. Below T_g, the expansion curves often have a slight curvature. The a_g reported are generally taken from the slopes of tangent lines drawn in the middle of the temperature range extending from 10 to 25 degrees below T_g.

cured with a broad range of hardeners suggests that the data spread seen in Fig. 5 may be due to real differences in the expansion behavior of the various cured resins. Another trend seen for this data was that for a given epoxy resin, higher functional curing agents often yielded cured systems with lower CLTEs above T_g.

The two systems containing D.E.R. 383 epoxy resin that do not fall on the line were cured with aminoethylpiperazine

(AEP, trifunctional) and p-menthane diamine. The low T_g of the p-menthane diamine–cured resin indicates that the system is very undercured and that the p-menthane diamine is behaving largely as a difunctional chain extender rather than as a tetrafunctional curing agent. Both of the last two systems were found to be quite tough. The AEP-cured system gave tensile elongations in excess of 10% and showed macroscopic shear banding before failure. The toughness of resins cured with AEP had always been assumed to be due to their low T_g's. The data of Fig. 5 also suggest that the toughness may be related to the low average functionality of this system, which allows the reaction to reach greater epoxide conversion at gelation than for higher functionality systems [38]. Going from a tetrafunctional to a trifunctional amine hardener more than doubles the network mesh size [73]. In a similar manner, systems with a low average functionality such as the CET resins, which were known to be unusually tough, were expected to have higher thermal expansion coefficients above their T_g than conventional resins with similar T_g's. This was found to be the case. As shown in Fig. 7, CLTE measurements on cured Tactix 695 and XUS 19020.00 type epoxy resins show that although their glassy CLTEs (a_g) are similar to those of most cured epoxy resins, their rubbery CLTEs (a_r) are larger than those of other resin systems processing similar T_g's. This increased rubbery expansion may be the result of greater fractional free volume, which results in increased segmental motion.

B. Thermoplastic Systems

Glassy and rubbery CLTEs for a variety of thermoplastics are plotted versus their T_g's in Fig. 8. The data points for these thermoplastics fall in the same region as those for the thermosetting epoxy resins. The rubbery expansion coefficients show no correlation with T_g. This may be due to the smaller temperature range involved, or it may be typical of uncrosslinked systems.

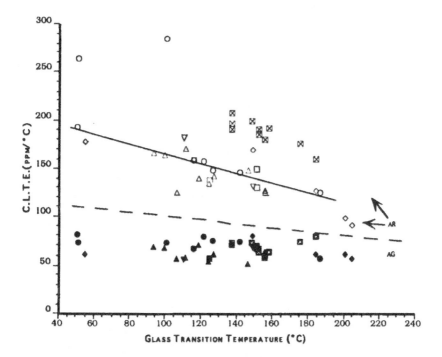

Figure 7 The thermal expansion of CET-type resins superimposed on the data of Fig. 5. Hollow symbols denote a_r (CLTE at temperatures $> T_g$), solid symbols denote a_g (CLTE at temperatures $< T_g$); ⊠ and ▨ denote CET-type resins (e.g., Tactix 695).

The theory of Gibbs and DiMarzio was used to calculate the energy barriers to segmental flow for polymers via statistical mechanics [C. B. Arends, internal communication of The Dow Chemical Company]. From this it was determined that the ratio of intra- to intermolecular barrier heights was the most pertinent measure of chain stiffness. When V_f/V_g was calculated using Eq. (1), given earlier, it was seen that the results for these nine thermoplastic polymers fell in the same order as those of the more complicated energy barrier calculations. Thus, thermal expansion coefficients appear to provide

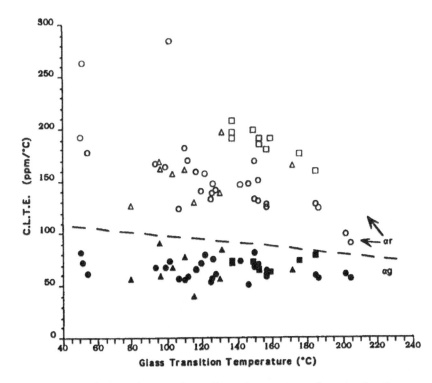

Figure 8 Thermal expansion data for various thermoplastics superimposed on Fig. 7. Hollow symbols denote a_r (CLTE at temperatures $> T_g$), solid symbols denote a_g (CLTE at temperatures $< T_g$); $\triangle$, thermoplastics; $\square$, CET-type resins; $\bigcirc$, other epoxy resins.

a useful way of ranking the chain (or perhaps network) stiffness of different polymers.

In summary, unusually tough epoxy resin systems have been obtained from single-phase resins such as Tactix 695 epoxy resin or experimental product XUS 19020.00 that are loosely crosslinked, have relatively stiff segments in the polymer chain, and may provide greater homogeneity of the cured epoxy network. At a given T_g, these epoxy systems show unusually high differences between their rubbery and glassy thermal expansion coefficients. It is hypothesized that these resins are tougher because either (1) a greater number of activated chain segments exist in the glassy state or (2) applied stress is more effective at increasing the number of activated chain segments.

XII. RESIDUAL STRESS MINIMIZATION

Residual stress minimization in composite materials is very important for many applications. There is a continuing demand for low stress compounds for use in semiconductor encapsulation and tooling. Stress in semiconductor encapsulation can arise from the mismatch of thermal expansion between the potting compound and the metal, ceramic, silicon, and so on, of the electronic device. There seems to be little likelihood of reducing these residual thermal stresses by significantly decreasing the glassy CLTE of the matrix resin, since the glassy CLTEs are similar for a wide range of polymer types. It was previously believed that by using a resin with the lowest T_g possible for the application, stress buildup would be minimized, since stress was reported to build only as the temperature is lowered from T_g [74]. However, Plepys and Farris [75], among others, showed that for three dimensionally constrained resins, thermal and polymerization shrinkage stresses can be large at temperatures above T_g, but may be reduced by cure schedule modification.

Residual thermal stresses in fiber-reinforced composites may affect the matrix-dominated properties of the composite such as solvent resistance, strengths transverse to the fiber direction, and compressive strengths. Since graphite and Kevlar fibers have two expansion coefficients each, radial and longitudinal, composites made from these fibers can never be stress free. The matrix thermal expansion coefficient required to eliminate the longitudinal stress while maintaining a compressive radial stress is much lower than is obtainable from cured epoxy resins [76]. However, the deleterious effects of such stresses may be decreased by using tough matrix resins such as CET resins. It should be kept in mind though that the use of unusually tough cured resins in composites may shift the failure site to the fiber/matrix interphase.

The effects of residual stress may also be mitigated by modifying the fiber/matrix interphase. Nairn [77] writes that residual stress in composites can be lowered by the interphase only if the interphase lowers the temperature for the onset of thermal stresses. He recommends that an amorphous thermoplastic with a T_g lower than the temperature for the onset of thermal stresses in the matrix be used (as a fiber sizing) to reduce matrix thermal stresses. Of course, this approach is limited by the increase of residual thermal stresses in the interphase [76,77].

XIII. ENVIRONMENTAL DEGRADATION

Thermal-oxidative, photooxidative, and thermal degradation of polymers can result in brittle surface layers that can initiate macroscopic failure of polymers [78,79], even those that have been fiber reinforced. Consequently, the degradation behavior of polymers should be understood for the environments in which they spend their service lives. Early screening of polymers for environmental degradation is appropriate, as these results may totally destroy the utility of toughness studies made on unaged polymers.

XIV. CONCLUSIONS

1. Dissolved rubbers serve to toughen cured epoxy resins through a yield stress reduction related to a reduction in the glass transition temperature of the system.
2. Rubbers that form secondary phases toughen by reducing residual hydrostatic tensile stresses. This is believed to occur through the imposition of a residual compressive

stress caused by thermal expansion mismatches and by stress-induced cavitation, during stressing, which relieves the triaxial stresses.

3. Epoxy resins containing insoluble preformed rubber dispersions can toughen without glass transition temperature reductions.

4. Unusually tough epoxy resin systems have been obtained from single-phase resins such as Tactix 695 epoxy resin or experimental product XUS 19020.00 that are loosely cross-linked, have relatively stiff segments in the polymer chain, and may provide greater homogeneity of the cured epoxy network.

5. At a given T_g, epoxy systems showing the greatest difference between their rubbery and glassy thermal expansion coefficients are likely to be tougher because either a greater number of activated chain segments exist in the glassy state or applied stress is more effective at increasing the number of activated chain segments.

NOTE: The information presented in this paper is presented in good faith, but no warranty, express or implied, is given, nor is freedom from any patent to be inferred.

ACKNOWLEDGMENTS

The following people have contributed to portions of this work: C. B. Arends, J. A. Clarke, L. L. Walker, C. E. Allen, T. Hammons, H.-J. Sue, P. M. Puckett, S. J. Maynard. Their assistance is gratefully acknowledged.

REFERENCES

1. L. E. Nielsen, *Mechanical Properties of Polymers and Composites*, Vol. 2, Marcel Dekker, New York, 1974, p. 299.

2. A. J. Kinloch and R. J. Young, *Fracture Behavior of Polymers*, Applied Science Publishers, New York, 1983.

3. R. N. Haward, *The Physics of Glassy Polymers*, Applied Science Publishers, London, 1973.

4. R. A. Bubeck, S. E. Bales, and H. D. Lee, *Polym. Eng. Sci.* 24:1142 (1984).

5. Michael Fischer, *Macromolecules: Synthesis, Order and Advanced Properties*, Advances in Polymer Science, Vol. 100, Springer-Verlag, 1991, p. 334.

6. V. T. Truong and B. C. Ennis, *Polym. Eng. Sci. 31*:548 (1991).

7. C. B. Bucknall and T. Yoshii, *Br. Polym. J. 10*:53 (1978).

8. C. K. Riew, E. H. Rowe, and A. R. Siebert, *Toughness and Brittleness of Plastics*, American Chemical Society, Advances in Chemistry Series No. 154, p. 326.

9. S. C. Kunz, J. A. Sayre, and R. A. Assink, *Polymer 23*:1897 (1982).

10. W. D. Bascom, R. Y. Ting, R. J. Moulton, C. K. Riew, and A. R. Siebert, *J. Mater. Sci. 16*:2657 (1981).

11. K. Yamanaka, Y. Takagi, and T. Inoue, *Polymer 60*:1839 (1989).

12. L. T. Manzione, J. K. Gillham, and C. A. McPherson, *J. Appl. Polym. Sci. 26*:889 (1981).

13. A. C. Grillet, J. Galy, and J.-P. Pascault, *Polymer 33*:34 (1992).

14. D. Verchere, J. P. Pascault, H. Sautereau, S. M. Moschair, C. C. Riccardi, and R. J. J. Williams, *J. Appl. Polym. Sci. 42*:701 (1991).

15. E. Butta, G. Levita, A. Marchetti, and A. Lazzeri, *Polym. Sci. Eng. 26*:63 (1986).

16. T. K. Chen and Y. H. Jan, *J. Mater. Sci. 27*:111–121 (1992).

17. A. C. Grillet, J. Galy, and J. P. Pascault, *Polymer 33*:34 (1992).

18. S. Wu, *J. Appl. Polym. Sci. 35*:549 (1988).

19. D. K. Hoffman, G. C. Kolb, C. B. Arends, and M. G. Stevens, *Polym. Preprints 26*:232 American Chemical Society (April 1985).

20. D. K. Hoffman and G. C. Kolb, *Sci. Adv. Mater. Proce. Eng. Ser. 35*:381 (1990).

21. D. K. Hoffman and C. B. Arends, U.S. Patent 4,708,996 (1987).

22. H.-J. Sue, *Polym. Eng. Sci. 31*:275 (1991).

23. H.-J. Sue, E. I. Garcia-Meitin, B. L. Burton, and C. C. Garrison, *J. Polym. Sci. Polym. Phys. Ed. 29*:1623 (1991).

24. D. E. Henton, D. M. Pickelman, C. B. Arends, V. E. Meyer, U.S. Patent 4,778,851 (1988); P. C. Yang and D. M. Pickelman, U.S. Patent 5,079,294 (1992).

25. H. George Waddill, *13th National SAMPE Technical Conference*, *11*:282 (1979).

26. Y. Huang and A. J. Kinloch, *Polymer 33*:1330 (1992).

27. T. Fukui, Y. Kikuchi, and T. Inoue, *Polymer 32*:2367 (1991).

28. R. A. Pearson and A. F. Yee, *J. Mater. Sci. 24*:2571 (1989).

29. J. B. Enns and J. K. Gillham, *J. Appl. Polym. Sci. 28*:2831 (1983).

30. K. P. Pang and J. K. Gillham, *J. Appl. Polym. Sci. 37*:1969 (1989).

31. K. P. Pang, Ph.D. dissertation, Princeton University, January 1989.

32. X. Wang and J. K. Gillham, *J. Coat. Technol. 64*:37 (1992).

33. K. P. Pang and J. K. Gillham, *Poly. Mater. Sci. Eng. 56*:435 (1987).

34. A. Noordam, J. J. M. H. Wintraecken, and G. Walton, *Crosslinked Epoxies*, Walter de Gruyter, New York, 1987, pp. 373–389.

35. S. M. Singer, J. Jow, J. D. DeLong, M. C. Hawley, *SAMPE Quart. 20*:14 (1989).

36. C.-S. Wu, Ph.D. dissertation, University of Illinois, 1985.

37. F.-M. Kong, D. M. Hoffman, and R. J. Morgan, *Org. Coatings Appl. Polym. Sci. 46*:599 (1981).

38. P. J. Flory, *Principles of Polymer Chemistry*, Cornell University Press, Ithaca, NY, 1953.

39. F. Bueche, *J. Polym. Sci. 24*:189 (1957).

40. S. C. Misra, J. A. Manson, and L. H. Sperling, *ACS Symposium Series No. 114*:137 (1979).

41. S. L. Kim, J. A. Manson, and S. C. Misra, *ACS Symposium Series No. 114*:183 (1979).

42. Wen-Li Wu, J.-T. Hu, and D. L. Hunston, *Polym. Eng. Sci. 30*:835 (1990).

43. E. F. Oleinik, *Advances in Polymer Science*, Vol. 80, Springer-Verlag, Berlin, 1986, pp. 50–99.

44. Riccardi and Williams, *Polymer 33*:44 (1992).

45. J. Mijovic and J. A. Koutsky, *Polymer 20*:1095 (1979).

46. J. Mijovic and L. Tsay, *Polymer 22*:902 (1981).

47. F. N. Kelly, B. J. Swetlin, and D. Trainor, *Macromolecules: 27th International Symposium* (H. Benoit and P. Rempp, Eds.), Strasbourg, France, 1981, pp. 275–288.
48. S. T. Wellinghoff and E. Baer, *J. Appl. Polym. Sci.* 22:2025 (1978).
49. J. P. Bell, *Org. Coatings Appl. Polym. Sci.* 46:585–591 (1982).
50. M. J. Galante, A. Vasquez, and R. J. J. Williams, *Polym. Bull.* 27:9 (1991).
51. B. L. Burton, *International SAMPE Technical Conference Series 18*:124 (1986).
52. J. L. Bertram, L. L. Walker, J. R. Berman, and J. A. Clarke, U.S. Patent 4,594,291 (1986).
53. J. L. Bertram, V. I. W. Stuart, L. L. Walker, U.S. Patent 4,725,652 (1988).
54. B. L. Burton, *International SAMPE Technical Conference Series 19*:653 (1987).
55. R. J. Morgan, *Epoxy Resins and Composites*, Vol. I (K. Dusek, Ed.), Elsevier, New York, 1985, p. 38.
56. B. L. Burton and C. A. Swartz, ACS Polym. Mater. Sci. Eng. Proc. 57:201–205 (1987).
57. D. J. Hourston and M. Lane, *Polymer 33*:1379 (1992).
58. K. Yamanaka and T. Inoue, *Polymer 30*:662 (1989).
59. C. B. Bucknall and A. H. Gilbert, *Polymer 30*:213 (1989).
60. H. Jabloner, B. J. Swetlin, and S. G. Chu, U.S. Patent 4,656,207 (1987); S. G. Chu, H. Jabloner, B. J. Swetlin, U.S. Patent 4,656,208 (1987); J. K. Bard, U.S. Patent 4,680,076 (1987).
61. K. R. Hirschbuehler and B. A. Stern, U.S. Patent 4,539,253 (1985).
62. R. E. Evans, J. E. Masters, and J. L. Courter, *Advanced Composites Conference Proceedings* December 1985, Dearborn, MI, ASM, pp. 249–257.
63. J. E. Masters, *Sixth International Conference on Composite Materials*, The Metallurgical Society, Vol. 3, 1987, p. 3.96.
64. L. E. Nielsen, *Mechanical Properties of Polymers and Composites*, Vol. 2. Marcel Dekker, New York, 1974, pp. 434–438.
65. T. T. Wang and T. K. Kwei, *J. Polym. Sci. A2.* 7:889 (1969).
66. R. F. Boyer, *Rubber Chem. Technol. 36*:1303 (1963).
67. R. Simha and R. F. Boyer, *J. Chem. Phys. 37*:1003 (1963).

68. R. Simha and R. F. Boyer, *J. Polym. Sci., Poly. Lett. Ed. 11*:33 (1973).
69. R. N. Haward, *The Physics of Glassy Polymers*, Applied Science Publishers, London, 1973.
70. J. C. Bauwens, *Polymer 21*:699 (1980).
71. M. R. Tant and G. L. Wilkes, *Polym. Eng. Sci. 21*:874 (1981).
72. *Thermal Characterization of Polymeric Materials* (E.A. Turi, Ed.), Academic Press, New York, 1981.
73. M. Fisher, F. Lohse, and R. Schmid, *Makromol. Chem. 181*:1251 (1980).
74. M. Shimbo, M. Ochi, and K. Arai, *J. Coatings Technol. 56*:45 (1984).
75. A. R. Plepys and R. J. Farris, *Polymer 31*:1932 (1990).
76. J. A. Nairn, *Polym. Compos. 6*:123 (1985).
77. J. A. Nairn and P. Zoller, *J. Mater. Sci. 20*:355 (1985).
78. B. L. Burton, *Eng. Plast. 3*:89 (1990).
79. B. L. Burton, *J. Appl. Polym. Sci. 47*:1821 (1993).

11
Toughness in Thermoset Coatings

ROBERT A. DuBOIS, DAVID S. WANG,
DAVID SHEIH* The Dow Chemical
Company, Freeport, Texas

I. INTRODUCTION

Polymer toughness has attracted a great deal of creative and resourceful attention from researchers for some time. The bulk of that attention has been directed at composites and bulk materials, and very little has been directed at coatings. Despite this drawback coatings researchers have still managed to utilize some of the work directed at composites to help in their own work. Fortunately, review articles are available that demonstrate applicability of the concepts of toughness to coatings [1–5].

Polymer coatings are used to cover a wide variety of substrates such as wood, paper, plastics, metal, and concrete for purposes that include decoration, weatherability, corrosion protection, and wear. Although the concepts of toughness can come into play to some extent in many of these applications, it

---

*Retired

†*Current affiliation:* Texas A&M University, College Station, Texas

is in connection with coatings on metal sheet or on objects formed from metal sheet that they have been most extensively studied; thus, that is to which we will confine our discussion.

The term toughness is rarely used in connection with coatings. The various combinations of flexibility and hardness that one typically associates with toughness are commonly expressed by the terms formability or postformability, fabricability, processability, impact resistance, and chip resistance. The requirements for toughness in coatings of metal articles that have been manufactured from flat stock can come either at some points in the manufacturing process, or in the end use of the finished article, or both. Generally, formability or postformability, fabricability, and processability refer to toughness requirements in the manufacturing processes, chip resistance refers to end use requirements such as in autos or furniture or building materials or appliance casings, and impact resistance refers to all of these. In the manufacturing processes, coatings can be applied (1) to the flat stock before various stamping and forming processes, (2) to the fully formed article, or (3) to the partially formed article, which then undergoes additional deformation often in finishing operations. Because the first alternative is generally the simplest and most economical, there is constant pressure toward developing coatings that will survive more severe stamping and forming operations [6]. A fuller understanding of the fundamental factors involved in coating formability can result in new products that can capture a larger share of the coatings market.

Coatings subjected to extreme mechanical abuses undergo a variety of stresses, including elongation, compression, impact, and abrasion. An illustration of how severe the demands can be is in the making of two-piece can bodies by the draw redraw (DRD) process. In this case cans are formed by punching a metal disc with a plunger into a die to give a shallow, broad "fishcan" as a first step followed by another plunger of smaller diameter into a deeper die to give the final shape. Such coatings undergo about 200% elongation in some areas and severe compression in others. Although it would appear from this example that the requirements for toughness in coatings are much more severe than those for composites, at least in terms of elongation and adhesion, it should be noted that the differences between thin-film coatings and bulk composites are so great as not to allow such a conclusion.

There are perhaps two major differences between coatings and composites that warrant further discussion: first, the presence or absence of a substrate and second, thin-film form versus bulk or thick-film form. Plastic materials can be toughened by altering the failure mechanism [7], such as the formation of crazes, voids, cracks, shear bands, and so on. However, this type of approach to toughening coatings provides little or no improvement in coating performance. Because of the nature of thin films, even microscale damages in the coatings could be serious enough to cause failure. Regardless of the types of failure, crazes, microcracks, major cracks, etc. coatings are rated "fail" as long as any of these damages are present. A change in toughening mechanism from one type of failure to another would not upgrade the coating from "fail" to "pass." The so-called tough coatings are those that "pass" severe deformation without showing failure. Therefore, coating toughness is considered to be the capability to withstand the deformation, rather than to resist propagation of cracks. It is preferable to describe coating toughness in terms of the stress–strain relationship rather than in terms of the fracture energy. In composites, increased toughness is often achieved by use of rubberized phases that minimize crack tip initiation and propagation. In coatings, we are not aware of any advantages to the use of rubberized-phase systems for coating formability or toughness; and, perhaps for that reason, the overwhelming majority of formable coatings are homogeneous systems. With respect to the form of a material, materials under stress that shatter or fail in brittle fashion when in bulk form often exhibit considerable flexibility when in the form of a thin film [8].

The purpose of this chapter is to review some of the relationships between structure and coating toughness of thermoset polymers and to show how toughness is revealed in thermomechanical measurements. Although coating structure is discussed in terms of the composite parts of structure—namely, backbone flexibility, crosslinking, and adhesion to substrate—the effects on toughness are considered in relation to more than just the separate contributions of each of the composites; the effects are considered particularly in relation to the interactions and interdependencies of the composites. We also discuss some of the approaches that have been taken to improve coating toughness.

II. COATING T_g

As a first approximation, it is generally recognized for thermoset polymers that T_g correlates strongly in some fashion with toughness. The ways to improve toughness in a system often cause a reduction in T_g. In the coatings field also, the significance of T_g in relation to toughness is well recognized. In fact, some manufacturers intentionally operate their stamping and forming equipment at temperatures near the T_g of the precoated metal stock in order to improve the ability of coating to survive the operation. The technique, referred to as *warm forming* or T_g *Forming* [9–12], is an example of adjusting the temperature of the environment to match the T_g of the coating as opposed to the more common approach of adjusting the T_g of the coating to match ambient temperature.

The T_g or flexibility of a thermoset is sensitive to variations in molecular structure of both the polymer and crosslinker components of the cured system and to crosslinking. Stutz et al. [13] used a model to describe the T_g in terms of three components: (1) the ideal backbone glass temperature, which depends on the structures of the backbone units and on the relative amounts of the units; (2) "the influence of end groups present, which lowers the glass temperature" [13] relative to the ideal backbone glass temperature; and (3) the effect of crosslinking, which increases the glass temperature relative to the ideal backbone temperature. Halary et al. [14] showed for a series of amine-cured epoxies that T_g decreases as backbone flexibility of the crosslinked system increases and increases as the crosslink density increases. Bellenger et al. [15] derived a model to calculate T_g's from a series of 40 different amine-crosslinked epoxies. The model is based on two components: one due to backbone structure and another due to crosslinking. Each of the chain segments in the network contributes to T_g in an additive fashion.

Structural modifications to the backbone that are designed to lower T_g are defined as examples of internal plasticization. Although T_g is also sensitive to external plasticization (i.e., plasticization induced by substances that are not chemically bound to the cured polymer structure), the subject has been reviewed elsewhere for coatings [16] and will not be considered here.

The strategies used to increase epoxy polymer flexibility by chemical modification are too numerous to list here. Some representative examples can be found in a few selected references [17–21]. Several workers in Dow Japan have shown that incorporation of flexible groups, dimer acid [22], or carboxyl-terminated polyamide made from dimer acid and ethylene diamine [23] into the backbone of an advanced epoxy resin results in a significant reduction in T_g of the cured coatings (which in turn is reflected by significant improvements in formability). In our own laboratory we conducted a study of the effect of systematic variations in the backbone structure of advanced epoxy resins on resin T_g, coating T_g, and finally coating formability [24–26]. The resin backbones used in this study consisted of various combinations of Bis A and alkylenedioxy-bridged diphenols. The alkylenedioxy chains served as the vehicle for structural variations. The fact that resin T_g's corresponded well with the values calculated from group contribution factors allowed us to, in effect, "dial in" a T_g for the epoxy resin. Because T_g's of the coatings paralleled those of the resin, at a given level of crosslinker and catalyst, we were even able to extend our ability to "dial in" a T_g to the cured coatings as well.

Crosslinking increases the T_g of a polymer system by restricting chain mobility. The effects of crosslinking on T_g are most often discussed in terms of crosslink density; however, Misra et al. [27] demonstrated that the distribution of crosslink density can also affect T_g. For a given average crosslink density the systems with a wide distribution of crosslink density exhibited lower T_g's than those with a narrow distribution. In other words, T_g is dominated by the longer segments between crosslinks. Other workers have also found that T_g responds less to increases in crosslinking for systems with flexible backbones than for those with stiffer backbones [28]. Bellenger et al. [15] found that the effects of chain structure and crosslinking contributed about 60% and 40% respectively to the T_g. They also found that chain stiffness due either to the presence of aromatic groups or to hindered rotation around the backbone axis (e.g., the isopropylidene or sulfone bridges in bisphenols) accounted for an increase in T_g of up to 60 to 110°C, and that meta-substituted aromatic rings contributed less to the T_g than the corresponding para isomers. We have observed similar effects for the cured epoxy coatings based on alkylenedioxy-bridged diphenols [25]; that is, T_g increased as crosslinking was increased but at a much lower slope for those resins with more flexible backbones. In studying the effect of stoichiometry on T_g of thermoset epoxy systems, Vallo et al. [29] have concluded that the effects are controlled by whether or not a network is "rigid" or "flexible." For a rigid network like that based on diglycidyl ether of bisphenol (DGEBA) the T_g can be predicted by considering only the concentration of elastic chains (crosslinking). For a flexible network the T_g depends more on backbone flexibility than on crosslinking. The difference was attributed to the greater facility of flexible chains to undergo relaxation within the chain structure as compared to rigid chains. The practical significance is evident—networks with flexible backbone structures can be crosslinked to the high levels often required for high solvent or chemical resistance without also becoming brittle.

From the preceding discussion it is clear that the combined effect of backbone flexibility and crosslinking on T_g is relatively complex. There is an interaction between the two such that one cannot always expect to be able to counteract the effect on T_g of increasing or decreasing backbone flexibility by proportionately increasing or decreasing crosslinking. Even if one could always do so, we will show below that good coating formability depends on more than T_g.

III. COATING FORMABILITY

Although backbone flexibility and crosslinking interact in their effect on the T_g of cured coatings, the effect of the interaction is one of degree not direction. The link between coating flexibility (T_g) and coating formability is more complex. In our work, coating formability, as determined by the impact wedge bend test and the reverse impact test and even the Erichsen fischan stamper test, exhibited maximum response over a range of coating T_g's of about 45 to 75°C. It is useful to note at this point that the industry standard for good formability in can stamping is polyvinyl chloride (PVC) organosol coatings, which exhibit T_g's of about 70°C, which is coincidentally within the optimum range of 45 to 75°C mentioned earlier for epoxy coatings; it should also be pointed out that the temperature the coating sees during the can-stamping operation is about 60°C [30]. According to one group of researchers automotive coatings may experience substantially higher temperatures as a result of stone impact (up to a 200°C rise in temperature) [31] than can coatings experience from stamping operations. Other workers have also recently reported a maximum in the response surface of formability with respect to T_g in the stamping and drawing operations of a variety of polymer systems similar to that we reported above [30,32]. In his attempts to find suitable replacements for PVC organosols, Hickling reported that the thermomechanical behavior of any replacement would have to be similar to that for PVC organosol [30]. Although the relationship between T_g and formability often works well, we have observed enough exceptions to indicate that additional factors are operative; thus, any attempt at attaining the proper T_g range should be viewed only as a first approximation or as a coarse adjustment.

In order for the relationship between T_g and coating formability discussed above to hold up, the extent of crosslinking must be held within some limits. It should not be so high as to make the coating brittle nor so low that the coating lacks sufficient cohesive strength. We have shown in our own work with epoxy resins based on alkylenedioxy-bridged diphenols and with a variety of polyamide-modified epoxy resins that as crosslinking is varied from low to high, coating formability goes through a maximum. Just above the optimum range the coatings are characterized by low methyl ethyl ketone (MEK) extractions and high MEK double rubs and by good T-Peel adhesion tests that show adhesive failure at the coating/substrate interface, and by formability tests that show failures occurring in a continuous fashion—all of which suggests good cohesive strength for the coatings. Just below the optimum range the coatings are characterized by high MEK extractions and low MEK double rubs and by low T-Peel adhesion tests that show cohesive failure, and by formability tests that show failures occurring in a spotty, erratic fashion—all of which indicates poor cohesive strength. At the optimum level of crosslinking the coatings are characterized by moderate to low MEK extractions and moderate to high MEK double rubs, and by good T-Peel adhesion tests that show adhesive failure, and by formability tests that show minimum failure occurring in either continuous fashion or spotty fashion or both.

It should be clear from the preceding discussion that comparisons of coating formability for different resin systems are valid only when they are made with reference to the extent of crosslinking for each of the systems. It is important to make such comparisons in areas of relatively flat response in order to minimize the error and to maximize the area of overlap between high formability and moderate to high crosslinking. Figures 1 and 2 show a response surface plot and contour plot, respectively, of formability in terms of percent

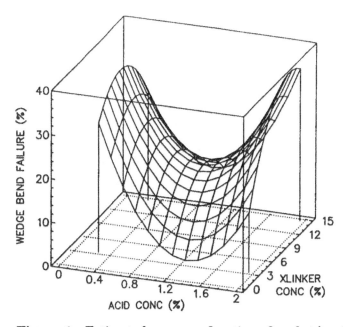

Figure 1 Estimated response function of wedge bend formability with respect to concentration of crosslinker and acid catalyst for a coating formulation of a standard Bis A–based solid epoxy resin, a resole crosslinker, and phosphoric acid cured 40 sec at 450°F.

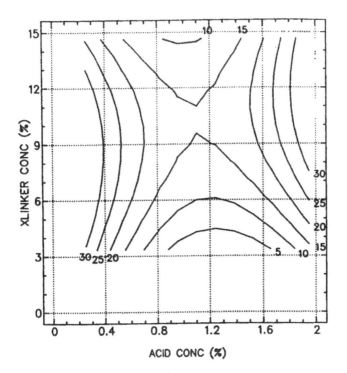

Figure 2 Contour plot for estimated response function of Fig. 1. The contour lines represent formability as measured by the impact wedge bend test in terms of percent failure of coating.

failure in the wedge bend test with respect to amount of acid catalyst and crosslinker for a standard Bis A–based epoxy resin; and Figs. 3 and 4 show two similar plots of relative extent of crosslinking in terms of percent solids extracted by MEK reflux for the same resin. Overlaying the two contour plots reveals two relatively small areas of overlapping optimum response (Fig. 5), that is, with wedge bend failures of 10% or less and MEK extractions of 20% or less. Figures 6 through 10 show response surface and contour plots for an epoxy resin system of comparable molecular weight but based on a flexibilized backbone. Notice the broader area of

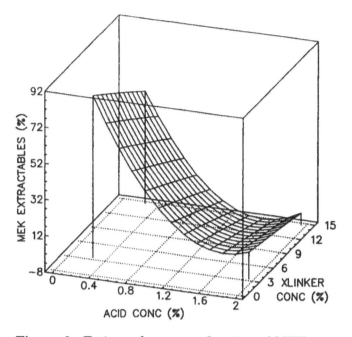

Figure 3 Estimated response function of MEK extractables with respect to concentration of crosslinker and acid catalyst for coating formulations of a standard Bis A–based solid epoxy resin, a resole crosslinker, and phosphoric acid cured 40 sec at 450°F.

maximum formability with respect to acid catalyst and crosslinker and the much broader area of overlapping optimum response. We have found that this behavior is typical for resins with flexible backbones. The effect has been demonstrated for amine-crosslinked epoxy composites as well; that is, networks made up of flexible chains exhibit high toughness over a much broader range of crosslinking density than do those made up of rigid chains [33]. This brings to mind our preceding discussion on T_g of coatings where we pointed out a lower slope in the response of T_g to increases in crosslinking for the resins with flexible backbones. Thus, T_g and formability of cured resins with flexible backbones are both

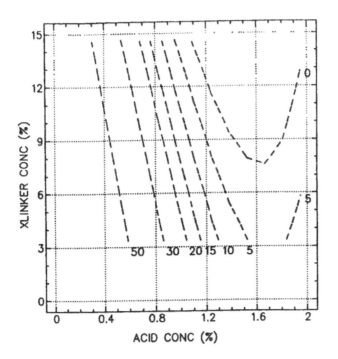

Figure 4　Contour plot for estimated response function of Fig. 3. The contour lines represent MEK extractables in terms of weight percent extracted from cured coating.

much more forgiving of crosslinking and thereby also of curing conditions than those with stiff backbones—this effect should allow more latitude in processing and formulating for resins with flexible backbones.

Even assuming that all the limitations of crosslinking have been properly considered as in the preceding discussion, yet another factor can confound the correlation between T_g and formability. We have found that a coating must exhibit some minimum level of adhesion to substrate or it will fail the various formability tests regardless of T_g. For the epoxy resin systems we have studied, that level is reflected by a T-Peel adhesion measurement of about 2 kg/5 mm [26]. The

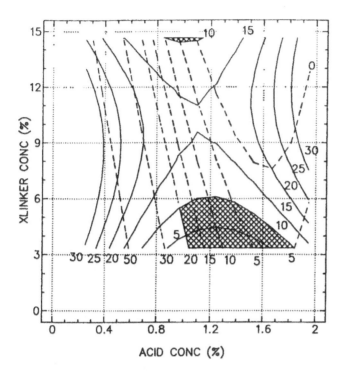

Figure 5 Overlay of contour plots of Figs. 2 and 4 for coatings of standard Bis A resin cured with a resole and phosphoric acid. Solid lines, formability in terms of percent failure in wedge bend test. Dashed lines, crosslinking in terms of percent MEK extractables. The crosshatch represents overlap of areas of maximum formability (<10% wedge band failure) and maximum crosslinking (<20% extractables).

model system that we used in our studies was well suited to reveal a direct correlation between T-Peel adhesion and a structural parameter of the resin, that is, concentration of aliphatic hydroxyl groups along the resin backbone. As the length of the alkylenedioxy bridging units was increased to increase backbone flexibility, the density of hydroxyl groups along the backbone decreased. Other workers have also re-

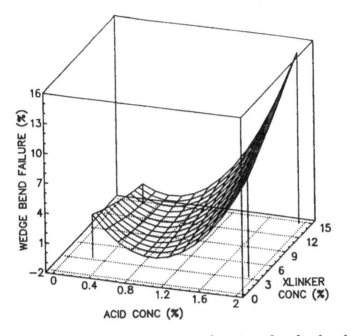

Figure 6 Estimated response function of wedge bend formability with respect to concentration of crosslinker and acid catalyst for coatings of a Bis A–based solid epoxy resin flexibilized by incorporation of carboxyl-terminated polyamide into the backbone and cured with a resole crosslinker and phosphoric acid at 40 sec and 450°F.

ported a direct correlation between adhesion and concentration of hydroxyl groups [34,35] and have even suggested a lower limit of hydroxyl concentration of 3.4% for minimally acceptable adhesion.

If the discussion to this point were meant to convey only one message, it would be that the factors controlling formability (backbone flexibility, crosslinking, and adhesion) are so strongly interactive that one can hardly vary one without also affecting the effect of the others on formability. For example, an attempt to improve formability by changing backbone

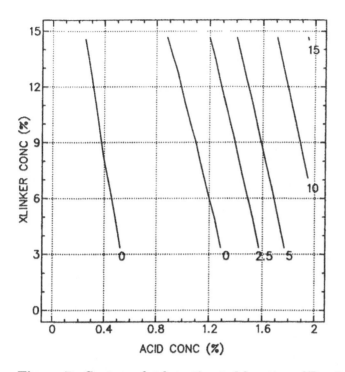

Figure 7   Contour plot for estimated function of Fig. 6. The contour lines represent formability as measured by the impact wedge bend test in terms of percent failure of coating.

structure in such a way as to reduce T_g could easily have an unintended effect on adhesion or on the position of optimum crosslinking that results ultimately in reduced formability. It is therefore naive to assume that operating on one of the factors in isolation from the others will produce the desired effect. Yet, even after having fulfilled all the requirements of T_g, and crosslinking, and adhesion, we still occasionally find some recalcitrant examples. Reconciling those examples requires a more detailed analysis of thermomechanical behavior than that required to obtain only the T_g.

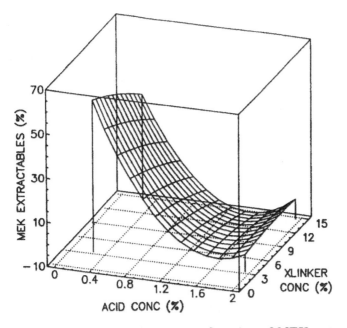

Figure 8   Estimated response function of MEK extractables with respect to concentration of crosslinker and acid catalyst for coatings of a Bis A–based solid epoxy resin flexibilized by incorporation of carboxyl-terminated polyamide into the backbone and cured with a resole crosslinker and phosphoric acid at 40 sec and 450°F.

IV.   THERMOMECHANICAL CONSIDERATIONS

Many examples in the literature show that flexibility as defined by T_g is not sufficient for coatings that must undergo impact or deformation. Hill relates that thermoplastic resins which show good flexibility and impact resistance are often characterized by low temperature peaks in the loss modulus and tan δ plots [4]. Polyester films cured with melamine formaldehyde are examples of such resins. They exhibit much better impact resistance than do comparable acrylic films

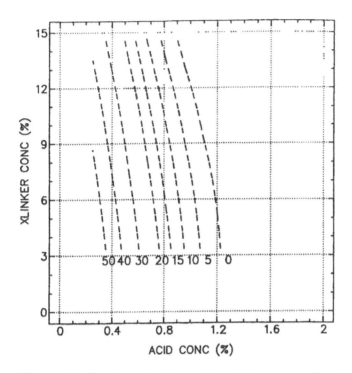

Figure 9 Contour plot for estimated response function of Fig. 8. The contour lines represent MEK extractables in terms of weight percent extracted from cured coating.

with the same T_g and crosslink density [5]. Hill further states that "For impact resistance, a binder must have some nonde-structive means of responding very rapidly to deformation . . . low temperature loss peaks are evidence for this response" [3]. Other workers show that a broad and intense tan δ peak is generally characteristic of good dissipation of impact energy [4,36,37].

Toughness or ductility has also been correlated by many workers in the composites area with a subglass secondary relaxation often referred to as the brittle–ductile transition, T_b [38,39]. Hill has argued that the concepts of brittle–ductile

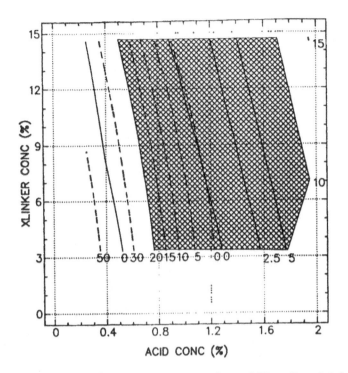

Figure 10 Overlay of contour plots of Figs. 7 and 9 for coatings of flexibilized resin cured with a resole and phosphoric acid. Solid lines, formability in terms of percent failure in wedge bend test. Dashed lines, crosslinking in terms of percent MEK extractables. The cross-hatch represents overlap of areas of maximum formability (<10% wedge bend failure) and maximum crosslinking (<20% extractables).

transition should be applicable to formable coatings [3,5]. In a study of formable PVC organosol coatings, Hickling has attempted to show that a wide separation between T_g and T_b (which he said is often seen as a broadening of the tan delta peak) is a prerequisite for good formability. We have applied dynamic mechanical thermal analysis (DMTA) techniques to determine the T_b, the T_g/T_b range, and the ultimate elongation of thermoset films and have related the results to the can-forming process [26]. The coating film was tested by superimposing a small dynamic deformation on a static pretension load. The ductile yielding point shifted toward lower temperature when the pretension load on the coating film increased. Figure 11 shows the typical tensile stress–strain curve of coatings under various testing temperature. When pretension stress increases from CC' to BB', the ductile yielding temperature that is detected from the sharp increase of loss tangent shifts from temperature (T3) down to lower temperature (T2). The T_b could be determined from the ductile yielding temperature under a maximum pretension load—above which the brittle fracture occurs. This testing method is found very useful in determining the T_b of thin (5 to 10 μm) coating films.

The conventional method of determining the ultimate elongation of coatings is by application of the Instron tensile test to free films of the coatings. Unfortunately, free films often contain flaws that can initiate cracks, which in turn can significantly reduce elongation at break. However, when the coating/substrate system is deformed, the strength of the substrate can provide much of the resistance to crack propagation through adhesive forces to the coating. Therefore, a tensile test of coating on a substrate can be considered as a tensile test of a flawless coating film. Van Krevelen [40] found a linear relationship between ultimate elongation and Poisson's ratio of polymers. Because Poisson's ratio correlates with the cumulative loss tangent, the loss tangent measured from DMTA could be applied to estimate the ultimate elongation of the coatings and thereby eliminate the problem associated with flaws. The estimated elongation is much greater than the Instron film tensile test value, and is closer to the true elongation of the coatings which are supported by a substrate. An example of the correlation between the cumulative loss tangent and the corresponding elongation of various coatings is shown in Fig. 12.

As mentioned earlier, we had found in our own work some exceptions to the rule of thumb that we had devised; that is, a

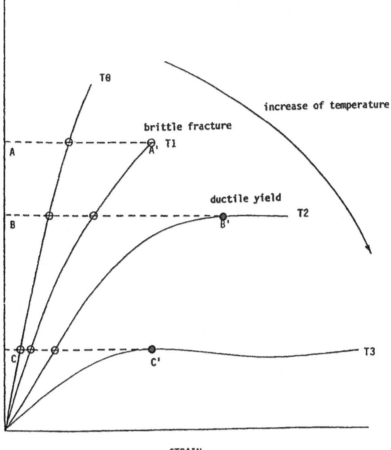

Figure 11   Typical tensile stress–strain curves of coating films at various temperatures.

LOSS TANGENT

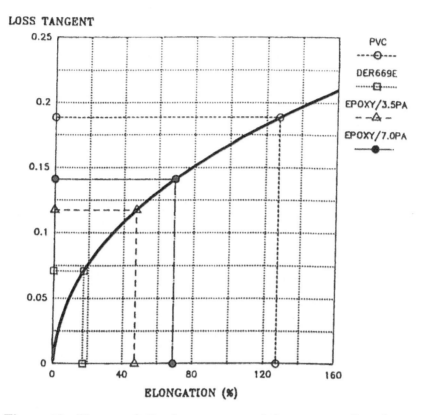

ELONGATION (%)

Figure 12   The cumulative loss tangent and the corresponding elongation of various coatings. ○, PVC; □, DER™ (trademark of The Dow Chemical Company) 669E epoxy resin; △, epoxy resin flexibilized with 3.5% polyamide; ●, epoxy resin flexibilized with 7.0% polyamide.

metal coating must have a T_g within the range of 45 to 75°C for good formability in applications involving stamping or drawing operations conducted under ambient conditions, provided the minimum requirements for adhesion and cross-linking have also been met. In one case a coating with a T_g of 72°C, just within the optimum range, still did not exhibit good formability; and conversely in another case a coating with a T_g of 78°C, just above the optimum range, nevertheless still exhibited very good formability. Although DMTA of the first coating revealed a temperature range of ductility (i.e., T_g/T_b) within that of PVC organosol coatings, it was extremely narrow (68 to 72°C) compared to the broad T_b/T_g range of about 50 to 70°C characteristic of PVC organosol coatings. On the other hand, the second coating exhibited a very broad T_b/T_g range (60 to 78°C) that extended well down into that of PVC organosol [26].

Coatings of a series of epoxy resins modified as described in Reference 23 by incorporation of polyamide oligomers into the resin backbone all performed significantly better in forma-

Table 1   Effects of Backbone Flexibilization with Polyamide and Bis F on Epoxy Resins Cured with 5% Methylon™ 75108 Resole and 0.75% H₃PO₄[a]

Advanced resin[b] (LER/Bis)[c]	Polyamide (%)[d]	T_b/T_g[e] (°C)	Wedge bend[f] (% Pass)
PVC Organosol[g]	—	74/52	100
Bis A/Bis A	0	100/65	84
Bis A/Bis A	3.5	75/57	92
Bis A/Bis A	7	85/52	100
Bis F/Bis A	3.5	78/43	100
Bis F/Bis A	7	70/42	100

[a]As described in Ref. 26.
[b]Molecular weight of epoxy resins, 13,000 to 16,000.
[c]The notation for an alternating copolymer made by advancing a liquid epoxy resin, LER, which is the diglycidyl ether of a bisphenol, with a bisphenol.
[d]Content of low molecular weight carboxyl-terminated polyamide of dimer acid and ethylenediamine.
[e]Glass transition and brittle–ductile transition determined on free film by DMTA, as described in Ref. 26.
[f]Formability test for coatings on metal sheet conducted by ASTM D3281-84.
[g]Industry standard for formable can coatings.

bility tests than a comparable standard Bis A–based epoxy resin despite T_g's for some of the coatings that were above the optimum range. Analysis by DMTA revealed much lower T_b's for the polyamide-modified resins than for the standard epoxy resin and T_b/T_g ranges that matched PVC organosol quite well (Table 1).

V. EFFECT OF MOLECULAR WEIGHT

Although the focus of this chapter is on the effect of the interaction between backbone flexibility, crosslinking, and adhesion, it is well worth reminding ourselves that molecular weight of the epoxy resin can also exert considerable influence on formability [41–44]. For example, the data in Table 2 clearly reveal an increase in toughness, expressed in terms of an impact resistance test, of a Bis A–based epoxy resin coating cured with a phenolic crosslinker as the molecular weight of the resin is increased. We had also found that the area of overlap in contour plots of maximum wedge bend formability

Table 2   Effect of Epoxy Resin Molecular Weight on Impact Resistance

Resin molecular weight	Impact resistance (in./lb)
2,600	<6
4,600	<6
6,800	<6
10,300	~6
12,000	20
18,000	60
27,500	>160
40,000	>160
54,000	>160
75,000	>160

and maximum crosslinking as expressed in terms of MEK extractables with respect to concentration of crosslinker and acid catalyst was much greater for a higher molecular weight resin than for a lower molecular weight resin. Unfortunately, substantial penalites are often associated with the use of the higher molecular weight resins in coatings—for example, solubility and compatibility problems and the need for high levels of solvent in order to attain reasonable viscosities [43].

There are methods available for handling high molecular weight resins in coating formulations, but they are usually difficult and economically undesirable. One way to circumvent the disadvantages of high molecular weight resins in coating formulations is to use lower molecular weight resins with reactive end groups that can polymerize in a linear fashion during the curing phase [45–49]. This technique has been applied to epoxy resin coatings. Epoxy resins of relatively low molecular weight were blended in coating formulations with Bis A-capped epoxy resins such that the two could react in a linear fashion to build molecular weight on the substrate during the bake cycle [50–52]. In this way one could achieve the good coating properties characteristic of high molecular weight resins with the relatively high solids formulations characteristic of low molecular weight resins.

VI. SUMMARY

For the field of polymeric coatings on metal sheet stock the concepts of toughness are expressed in terms of the ability of a coating to survive various combinations of impact and abrasion and of elongations of up to about 200%. Some of this ability is drawn from the strength of the metal itself through the intercession of adhesive bonds between coating and metal, and some of it derives from the inherent flexibility of a thin film compared to the same material in bulk form. The remainder of this ability is related (1) to a complex interaction between crosslinking, backbone flexibility, and adhesion, (2) to the way in which the interaction between crosslinking and chain flexibility affects T_g, and (3) to how closely matched are the T_g and the temperature range that the coating experiences during deformation.

The T_g of a flexible network is much less affected by crosslink density than is the T_g of a rigid network. Thus, flexible networks offer much wider latitude in selecting crosslink densities to satisfy other properties. As a first approximation, the formability of a coating as expressed in the wedge bend test is at a maximum when T_g of the coating is within a range of about 45 to 75°C, provided crosslink density and adhesion are appropriate. The exceptions to this rule can be reconciled by consideration of dynamic mechanical properties. Coatings that exhibit broad tan δ peaks or broad temperature ranges of ductility (T_g/T_b range) or other low-temperature loss peaks of sufficient magnitude exhibit good formability even when the T_g is up to 10°C above the optimum range; conversely, coatings that exhibit narrow tan δ peaks or narrow ranges of T_g/T_b can exhibit poor formability even if the T_g is well within the optimum range. Increasing crosslink density not only increases T_g but also narrows the T_g/T_b window and depresses low-temperature loss peaks; this process is reflected by losses in formability.

Note: The information in this chapter is presented in good faith, but no warranty, express or implied, is given, nor is freedom from any patent to be inferred.

REFERENCES

1. Z. Wicks, Coatings, *Encyclopedia of Polymer Science and Engineering Suppl.* Vol. 53–122. (J. I. Kroschwitz, Ed.), John Wiley, New York, 1990.
2. R. H. Good, Recent Advances in Metal Can Interior Coatings, *Food and Packaging Interactions, ACS Symposium Series No.*

365, American Chemical Society, Washington, DC, 1987, Chap. 17.

3. L. W. Hill, Mechanical Properties of Coatings, *Federation Series on Coatings Technology*, Federation of Societies for Coatings Technology, 1987, p. 21.

4. L. W. Hill, The Relationship Between Dynamic Mechanical Measurements and Coating Properties, *Adv. Org. Coatings Sci. Technol. Ser. 10*:31 (1988).

5. L. W. Hill, Structure/Property Relationships of Thermoset Coatings, *J. Coat. Technol. 64*:29 (1992).

6. K. K. Deodhar, Pre-finished Steel-Step into Hi-Tech Hi-Class Decorative Metal Finish (Coilcoating Technology), *Paintindia 40*:13 (1990).

7. A. C. Garg and Y. W. Mai, Failure Mechanisms in Toughened Epoxy Resins—A Review, *Compos. Sci. Technol. 31*:179 (1988).

8. G. C. Fettis and S. B. Downing, Scientific Approach to the Development of Coatings, *Adv. Org. Coatings Sci. Technol. Ser. 11*:127 (1989).

9. D. A. Cocuzzi et al., *Polym. Paint Colour J. 179*:870 (1989).

10. J. J. Obrzt, Glass Transition Forming: A Boost for Coil Coaters, *Iron Age*, August 4, 1980, p. 63.

11. C. Noll, Applications of the Glass Transition in Forming Precoated Metal, *Metal Finishing*, January 1979, p. 27.

12. J. Oka, Precoated Paint Films with Excellent Processability, *Netsu Shori [J. Jpn. Soc. Heat Treatment] 29*:160 (1989).

13. H. Stutz, K.-H. Illers, and J. Mertes, A Generalized Theory for the Glass Transition Temperature of Crosslinked and Uncrosslinked Polymers, *J. Polym. Sci., Part B: Polym. Phys., 28*:1483 (1990).

14. J. L. Halary, S. Cukierman, and L. Monnerie, Relationships Between Chemical Structure and Mechanical Properites in Epoxy Networks, *Bull. Soc. Chim. Belg. 98*:623 (1989).

15. V. Bellenger, J. Verdu, and E. Morel, Effect of Structure on Glass Transition Temperature of Amine Crosslinked Epoxies, *J. Polym. Sci., Part B: Polym. Phys. 25*:1219 (1987).

16. P. Nylen and E. Sunderland, *Modern Surface Coatings*, Wiley-Interscience, 1965, p. 337.

17. H. Lee and K. Neville, *Handbook of Epoxy Resins*, McGraw-Hill, New York, 1967, Chap. 16.

18. J. W. Lister, Epoxy Resin Surface-Coating Compositions, G.B. Patent 2,075,021, 1986.

19. J. V. Koleske, Low Viscosity Adducts of a Polycaprolactone Polyol and a Polyepoxide, U.S. Patent 4,629,779, 1986.

20. P. W. Kopf and C. N. Mariam, Lactone-Grafted Polyethers in Coatings for Deformable Substrates, EP Patent 135,916, 1985.

21. J. T. K. Wook, V. Ting, J. Evans, C. Ortiz, G. Carlson, and R. Marcinko, Water-Dispersable Epoxy-*g*-Acrylic Copolymer for Container Coating, *Epoxy Resin Chemistry*, Vol. II (R. S. Bauer, Ed.), ACS Symposium Series No. 221, American Chemical Society, Washington, DC, 1983, Chap. 15.

22. K. Ohba, Preparation of High-Molecular Weight Dimer Acid-Modified Epoxy Resins, Eur. Pat. Appl. EP 379,943, 1990.

23. K. Ohba, Polyamide Epoxy Ester Resin, Process for Preparation Thereof and Coating Composition, U.S. Patent 5,070,174, December, 1991.

24. R. A. Dubois and P. S. Sheih, Fundamental Studies of Epoxy Resins for Can and Coil Coatings: III. Effect of Bisphenol Structure on Resin and Coating Flexibility *J. Coatings Technol.* *64*:51 (1992).

25. R. A. Dubois and P. S. Sheih, Novel Epoxy Resins Based on Alkylenedioxydiphenols—Effect of Backbone Flexibility and Crosslinking on Flexibility of Can Coatings, *Polym. Mater. Sci. Eng.* *65*:325 (1991).

26. R. A. Dubois and D. S. Wang, Effect of Structure on Coating Performance Properties of Novel Alkylenedioxydiphenol Based Epoxy Resins, *Prog. Org. Coatings* *22*:161 (1993).

27. S. C. Misra, J. A. Manson, and L. H. Sperling, Effect of Cross-Link Density Distribution on the Engineering Behavior of Epoxies, *Epoxy Resin Chemistry*, ACS Symposium Series No. 114, 1979, American Chemical Society, Washington, DC, p. 137.

28. Y. G. Won, J. Galy, J.-P. Pascault, and J. Verdu, Prediction of the Glass Transition Temperature of Cycloaliphatic Amine-Epoxy Networks, *J. Polym. Sci., Part B: Polym. Phys.* *29*:981 (1991).

29. C. I. Vallo, P. M. Frontini, and R. J. J. Williams, The Glass Transition Temperature of Nonstoichiometric Epoxy-Amine Networks, *J. Polym. Sci., Part B: Polym. Phys.,* *29*:1503 (1991).

30. M. Hickling, PVC in Container Coatings and Approaches to Its Replacement, *Polym. Mater. Sci. Eng.* *65*:285 (1991).

31. A. T. Zehnder, A. C. Ramamurthy, S. J. Bless, and N. S. Brar, Temperature Rise in Automotive Paint Coatings Due to Stone Impact, *Polym. Mater. Sci. Eng.* 67:116 (1992).
32. F. Kenichi, Modern Techniques in Precoated Metal Manufacturing, *JETI.* 38:39 (1990).
33. E. Urbaczewski-Espuche, J. Galy, J.-F. Gerard, J.-P. Pascault, and H. Sautereau, Influence on Chain Flexibility and Crosslink Density on Mechanical Properties of Epoxy/Amine Networks, *Polym. Eng. Sci.* 31:1572 (1991).
34. J. W. Holubka, J. E. deVries, and R. A. Dickie, *Ind. Eng. Chem. Prod. Res. Dev.* 23:63 (1984).
35. W. Raudenbusch, Development of Epoxy Resin–Based Binders for Electrodeposition Coatings with High Corrosion Resistance, *Epoxy Resin Chemistry* (R. S. Bauer, Ed.), ACS Symposium Series No. 114, American Chemical Society, Washington, DC, 1979, p. 57.
36. Z. Wicks, *Research Challenges in Coatings Science*, Fall American Chemical Society Meeting No., 269 (1988).
37. R. J. Dick, R. A. Markle, B. A. Mayo and J. P. Pfau, *Advances in Organic Coatings and Paints*, Battelle Technical Inputs to Planning, Report No. 63, OAR89-473B, 1989.
38. S. Wu, *J. Appl. Polym. Sci.* 20:327 (1976).
39. S. Wu, Secondary Relaxation, Brittle–Ductile Transition Temperature, and Chain Structure, *J. Appl. Polym. Sci.* 46:619 (1992).
40. D. W. Van Krevelen, *Properties of Polymers*, 3rd ed., Elsevier, Amsterdam, 1990.
41. C. A. May, *Epoxy Resins: Chemistry and Technology*, 2nd ed., Marcel Dekker, 1988.
42. H. Lee and K. Neville, *Handbook of Epoxy Resins*, McGraw-Hill, New York, 1967.
43. P. S. Sheih and J. L. Massingill, Fundamental Studies of Epoxy Resins for Can and Coil Coatings: I. Adhesion to Tin Free Steel, *J. Coatings Technol.* 62:25 (1990).
44. J. L. Massingill, P. S. Sheih, R. C. Whiteside, D. E. Benton, and D. K. Morisse-Arnold, Fundamental Studies of Epoxy Resins for Can and Coil Coatings: II. Flexibility and Adhesion of Epoxy Resins, *J. Coatings Technol.* 62:31 (1990).

45. R. C. Whiteside, P. S. Sheih, and J. L. Massingill, High Performance Epoxy Resins for Container Coating Applications Based on In-Situ Advancement Technology, *J. Coatings Technol.* 62:61 (1990).
46. E. J. Kuckhoff and J. L. Massingill, Epoxy-Phenolic Coatings with Higher Solids and Faster Cure, Proc. 13th Water-Borne and Higher-Solids Coatings Symposium, University of Southern Mississippi, Hattiesburg, 1986, p. 446.
47. J. L. Massingill, High-Solids Epoxy-Based Coatings, *Chemtech* 18:236 (1988).
48. R. D. Athey, Jr., Review of Telechelic Polymer Synthesis, *J. Coatings Technol.* 54:47 (1982).
49. R. F. Storey, Proc. 12th Water-Borne and Higher-Solids Coatings Symposium, University of Southern Mississippi, Hattiesburg, 1985, pp. 59–72.
50. P. S. Sheih and R. C. Whiteside, A Novel Epoxy System for Low VOC Heat Converted Coating Applications, Proc. 15th International Conference in Organic Coatings Science and Technology, Athens, Greece, 1989.
51. P. H. Martin, Process for Coating Substrates with High Molecular Weight Epoxy Resins, U.S. Patent 4,322,456, 1982.
52. R. Koenig, Epoxy Resins of Controlled Conversion and a Process for Their Preparation, Pat. Appl. W086/01216, 1983.

Index